BIODIVERSITY AND ENVIRONMENT

BIODIVERSITY AND ENVIRONMENT

Edited by

S.K. Agarwal

Swarnlata Tiwari

P.S. Dubey

A.P.H. PUBLISHING CORPORATION

4435-36/7, ANSARI ROAD, DARYA GANJ

NEW DELHI-110 002

Published by
S.B. Nangia
A P H Publishing Corporation
4435-36/7, Ansari Road, Daryaganj
New Delhi 110002
Ph.: 23274050
E-mail : aphbooks@gmail.com

2026

Printed at
Balaji Offset
Navin Shahdara, Delhi 110032

KAMAL NATH
Minister
ENVIRONMENT AND FORESTS
INDIA

FOREWORD

The festschrift on "Biodiversity and Environment", edited by Dr. S.K. Agarwal, Dr. Mrs. Swarnlata Tiwari and Dr. P.S. Dubey is a timely reminder to us of the new threats the humankind faces for its own survival.

The Earth and its environment are the true laboratories of nature which keep constantly churning out processes of various kinds. There is a certain balance, a certain ambience to be maintained in nature if all designated processes are to come to their fruition. With the nuclear threat receding into the background for sometime now, the problem now is not that of how man survives with man, but that of how man survives with nature!

While science has enabled us to live a life of quality and comfort, it has also spawned enormous amounts of pollution. A salutary effect, however, of development of science and technology, has been the fact that it has enabled us to evaluate the impact of the acts of our plunder of the nature and its natural resources. Consumerism and pollution are the prices that man has had to pay for his greed. The pollutants degrade the nature and pollute the environment. No organism in the biological history survives for long if its natural environment became, in some way, unfit for it. It is, therefore, of vital importance that environmental researches are conducted into the present and future needs of the Society so that the humankinds' needs are well taken care of, with minimum plundering of natural resources. It is here that the value of scientific evaluation and environmental research assumes crucial and critical importance.

In an attempt to reach materialistic targets, man has chalked out an ambitious plan of rapid industrialisation and urbanisation. In this venture, he has degraded and destroyed the biodiversity built-up so meticulously by nature over millions of years, so much so, that the term Development has become synonymous with Degradation, and Progress with Pollution.

The festschrift deals with review and research papers on all aspects of biodiversity, impact of environmental changes on biodiversity, industrial, urban and indoor environment's impact, aerobiology, environmental legislation etc.

The book contains 26 review and research papers contributed by eminent academicians from various Universities and research institutions. These valuable papers provide a picture of the new trends in environmental research in relation to the present needs of the society. This book also is an attempt at highlighting the 'basic truth' which is that everything has a price to be paid. Its publication is an opportunity to present to the reading public a volume of valuable scientific studies.

I hope this volume will render yeomen service by not only serving as a reference book but also by spreading the message of environmental research and regulation far and wide, and, in disseminating the 'basic truth'.

(Kamal Nath)

B.R. YADAV
Minister
ENVIRONMENT AND FORESTS,
MINERAL RESOURCES DEPTT.

FOREWORD

It gives me great pleasure to write a foreword to this festschrift on *Biodiversity and Environment.*

This earth and its environment is the true laboratory of nature. Processes of many kinds are continuously occurring in it. Lining organisms in soil, air and water owe their continued survival to the balance of the physical and chemical conditions prevailing on earth and its environment. In view of these ecological factors mankind has to see that no such circumstances prevail which can force an abnormal and harmful situation resulting in peril to the natural system.

Environmental science designates everything in order to understand the mechanism of nature and anticipate its evolution, which leads to the idea of the *Quality of life*. The twentyfirst century is likely to pose before us serious challenges of meeting pollution hazards. Environmental degrading is enjoined with serious threats to the very existence of life on this planet. Environmental researches are, therefore, of utmost importance.

The book contains 26 review and research papers contributed by eminent academicians from various Universities and research institutions. These valuable papers provide a picture of the changing and new trends in environmental research in relation to present needs of the society.

I am grateful to the Editors of this book for choosing me to write for this publication. I offer my best wishes for their sincere efforts and hope that this publication would prove to be a valuable one.

BRYadav

(B.R. Yadav)

PREFACE

Man has always been fascinated by the diversity of life. Biodiversity entails all forms of biological entities inhabiting the Earth–including Prokaryotes and Eukaryotes–wild plants and animals, micro-organisms and macro-organisms, domesticated and wild animals, cultivated and wild plants, and even genetic material like seed and germplasm.

Biodiversity is the result of the evolutionary plasticity of living organisms which has increased geometrically through perhaps 2.5 billion years, proliferating by trial and error, controlled by natural selection, filled almost every one of the habitable ecological niche created in a likewise evolving world environment.

In an attempt to reach materialistic targets, man has chalked out an ambitious plan of rapid industrialization and urbanisation. In this venture, he has not only destroyed the biodiversity build-up meticulously by nature over millions of years, so much so, that development has become synonymous with Degradation, and Progress with Pollution.

This fastschrift contains 26 review and research papers on the newer areas of Biodiversity and Environment, which provide picture of the changing and new trends in environmental research. The papers in this book are mostly invited ones from the eminent academicians from various Universities and institutions. We are thankful to the authors for their cooperation and active participation. The authors themselves bear the responsibility for their interpretation, literature citation and scientific names.

It is hoped that the vast information contained here will help the students, teachers and researchers alike, in their persuits.

Editors

CONTENTS

CONTRIBUTORS

1. **Dr. S.K. Agarwal**
Head, P.G. Department of Botany,
Govt. Autonomous College,
Kota 324001

2. **Mr. Ram Prasad**
IFS, Forest Department,
Bhopal (M.P.)

3. **Dr. M. Oomachan**
Professor in Biosciences,
R.D. University,
Jabalpur 482001

4. **Dr. S.H. Raza**
Professor in Botany,
Osmania University,
Hyderabad 7

5. **Dr. D.M. Tripathi**
Hon. Allergologist,
Bombay Hospital,
Institute of Medical Sciences,
Bombay 20

6. **Dr. Swarnlata Tiwari**
Professor in Botany,
Kirodimal Govt. Science College,
Raigarh 496001 (M.P.)

7. **Dr. Miss Samidha Bansal**
Scientist,
M.P. Pollution Control Board,
E-5, Arera Colony,
Bhopal 462016

8. **Dr. J.P. Kaushik**
Associate Professor,
School of Studies in Botany,
Jiwaji University,
Gwalior 474011

9. **Dr. H.K. Goswami**
Professor in Genetics,
Bhopal University,
Bhopal 462026

10. **Dr. S.N. Singh**
Environmental Botany Laboratory,
National Botanical Research Institute,
Rana Pratap Marg,
Lucknow 226001

11. **Dr. Smt. Swati Shrivastava**
Govt. Gitanjali Girls College,
Bhopal (M.P.)

12. **Dr. M.P. Singh**
Principal,
Govt. College,
Narsingarh (M.P.)

13. **Dr. R.K. Srivastava**
Botany Department,
Govt. Autonomous College,
Jabalpur 482001 (M.P.)

14. **Dr. A.K. Girolkar**
Botany Department,
Kirodimal Govt. College,
Raigarh 496001 (M.P.)

15. **Dr. A.S. Khalatkar**
Professor in Botany,
Nagpur University,
Nagpur 440010

16. **Dr. S.V.S. Chauhan**
Professor in Botany,
RBS College,
Agra 282002

17. **Dr. Akhilesh Ayachi**
Botany Department,
Govt. Autonomous College,
Jabalpur 482001 (M.P.)

18. **Dr. R.P. Mishra**
Department of Biological Sciences,
R.D. University,
Jabalpur 482001 (M.P.)

19. **Dr. K.S. Verma**
Reader in Biological Sciences,
R.D. University,
Jabalpur 482001 (M.P.)

20. **Dr. Pramod Kumar**
Scientist,
Birbal Sahani Institute of Palaeobotany,
53, University Road,
Lucknow 226007

21. **Dr. R.K. Sharma**
Department of Law,
Govt. Autonomous College,
Kota 324001 (Raj.)

22. **Dr. S.K. Shringi**
Botany Department,
Govt. Autonomous College,
Kota 324001 (Raj.)

23. **Dr. Paramjit S. Jaswal**
Reader in Law,
Panjab University,
Chandigarh 160014

24. **Dr. I.C. Saxena**
Former Principal,
Law College,
Rajasthan University,
Jaipur

25. **Mr. A.K. Sharma**
Department of Law,
Govt. Autonomous College,
Kota 324001 (Raj.)

26. **Dr. B.K. Galchhania**
Department of Law,
Govt. Autonomous College,
Kota 324001 (Raj.)

27. **Dr. T. Bhattacharyya**
Professor in Law,
Rajasthan University,
Jaipur.

BIODIVERSITY

S.K. Agarwal

Man has always been fascinated by the diversity of life. Biodiversity is the new international buzzword. Perhaps it has not attracted as much attention as global warming and ozone depletion, but it has certainly been catapulled into the centrestage of world-wide environmental politics in the last handful of years. Some people are under the impression that it is just another fancy term for wildlife, but it is not. Biodiversity entails all forms of biological entities inhabiting the Earth – including procaryotes and eu-karyotes – wild plants and animals, micro-organisms, domesticated animals and cultivated plants, and even genetic material like seeds and germplasm (Kothari, 1992).

Biodiversity is the result of the evolutionary plasticity of living organisms, and increased geometrically through perhaps 2.5 billion years, proliferating by trial and error, controlled by natural selection, filling almost every one of the habitable ecological niches created in a likewise evolving world environment. Our palaeontologists tell us that the number of species existing have fluctuated widely – how widely we can never know with any certainly. The causes of these fluctuations have been subjects of much speculation – varying from gradual and normal climatic change to catastrophic changes even as great as the impact of a hypothetical gigantic meteor or asteroid with the terrestrial globe (Fosberg, 1986).

Organic evolution, during its 2.5 billion years, has probably produced from time to time locally or regionally dominant species or groups of closely related species, but none that seriously lessened the almost explosive increase of organic diversity. In the last 100,000 or more years the primate mammals produced a species *Homo sapiens*, with a highly developed nervous system centred on an immensely complex and functional enlarged brain. For the first millenia of its history this species doubtless behaved as a normal member of the faunas of the ecosystems in which it occurred.It did gradually increase its geographic range over wide reaches of the Old

World continents, perhaps starting from Africa, where the homonid line apparently appeared first and slowly developed – at least so the oldest available homonid fossils tell us (Fosberg, 1986).

At the same time during the last few tens of millenia, this species, with its very effective brain, erect posture and thumb-opposed hands, began to exert a noticeable effect on its environment. It then started to consciously its environment to better suit its own requirements (Fosberg, 1986).

This development was at first so slow that fossil traces of it are almost imperceptible. Weapons and tools were invented and fire was captured and put to man's use. This was probably man's first major means of environmental modification. Characteristically foreshadowing the future, its effects were largely destructive. Its advantageous aspects were provision of warmth to expand man's habitat into colder regions, and to increase his food supply by rendering more foods edible by cooking. By use in driving game it further made available more food. A side benefit was to keep dangerous carnivores at bay.

But the unintended consequences of the use was to vastly increase the incidence of wild-fire with consequent greatly increased change of forest in drier or more seasonal climatic regions into savanna and grassland. This was the first great step in man's changing the face of the earth.

At first, this process may have, by increasing the relatively tiny areas of savanna, somewhat augmented the natural diversity of habitats and perhaps assisted, or permitted the evolution of organisms to fit this greater habitat diversity.

This effect of great increase in open spaces in an otherwise largely forested world, and the increase in number of ungulates to occupy and exploit it, was doubtless out of all proportions to man's numbers at the time.

Unquestionably as numbers of human beings increased, their effect in environmental modification increased likewise. In the Old World such favourable habitats as Southern China, the Indus Valley, Mesopotamia, the Nile Valley and the Eastern Mediterranean countries became centres of population increase. Resources were used, at first moderately, but as people became numerous, they were overused, and environmental degradation set in. The universal process of diversification seems to have been reversed. The *Homosapiens* has established world-wide dominance and is not just threatening, but destroying the marvellous diversity that has evolved over two and a half billion years (Fosberg, 1986).

Biodiversity comprises every form of life from the tiniest microbes to the mightiest beasts and the gigantic trees. The biodiversity exists at three different levels. These are : (i) Species diversity., which embraces the variety of living organisms on earth, (ii) Genetic diversity, which is concerned with the variation in genes within a particular species, and (iii) Ecosystem diversity, which is related to the variety of habitats.

Species diversity is reflected by morphological, physiological and genetic features. India is home of about 2,00,000 species of living organisms. Of the total species described, several are endemic to India.

Genetic diversity is the result not only of the number of species present in a given area, but also of successional changes from colonization of gaps by pioneer species to mature climax. It comprises genetic or other variations within a species – ecotypes, land races, horticultural varieties, cultivates etc., all within one biological species. We have a long tradition of domestic animals breeds, bred for specific traits. These include cattle, goats and sheep as well as horses and pigeons for sport. With emphasis on increasing milk yield through cross-breeding, some of the original cattle breeds are in the state of extinction. Similarly, high yielding hybrid plant varieties are to be replaced by locally adopted domestic varieties. This is a real challenge to our agriculture. India has been the centre for origin of most genetic stock of plant and animal species.

Ecosystem diversity is reflected in our diverse biogeographic zones such as lakes, deserts, coasts and estuaries. Some of these ecosystems are fragile – wetlands, mangroves and coral reefs. They are under constant pressure. The conservation of these ecosystems is a major challenge.

The biological movement towards complexity is not a new phenomenon. The beginning of life on this planet, as well as its subsequent evolution, is documented in the fossil record; here, biological evolution has been demonstrated to involve a hierarchical ascendency to more and more complex forms of living beings. This complexity increases not only in life forms but also in ecosystems. We humans are but one expression of that process. Any reduction of the conditions or options allowing progress along this path will stifle the expression of life on it. Observations confirm the concern that the actions of modern society are contributing to such a reduction of options.

Today, we rapidly remove the conditions necessary for the continued existence of genetic, ecological and cultural information, as these have been seen to be unimportant to modern society. These losses have now reached critical conditions and are beginning to be addressed as biodiversity loss (Senanayake, 1993).

Biodiversity has never been a focus of world-wide attention. The rich and the powerful in the global community have just realised its enormous economic potential. For the first time the question of biodiversity is not merely a matter of aesthetic or outdoor recreation for the elite but also a matter of money. They, therefore, see conservation of biodiversity as something that really matters (Gadgil, 1992).

The traditional diversity was bred to meet diverse human needs of nutrition, taste, colour, ritual, smell and to resist drought, flood and pests. It provided several kinds of insurance against crop failure to the farmer. Modern hybrids on the other hand, while substantially increasing the grain yield and monetary profits, have forced the farmers to look elsewhere for their other daily needs (especially fodder), and left them depend on the vagaries of markets, government and private corporations.

Indian cosmology places the estimates of biodiversity of living organisms at 84 lakh "yonies". Twenty years ago biologists thought this estimate to be too high. They had described 15 lakh species; the total was believed to be less than 50 lakh. However, when they began looking carefully at the humid tropical forests, they discovered species of trees differing from one valley to the next, and hundreds of different species of insects in the canopy of a single tree. The total number of species is now estimated to be as high as five crore, and the population of each species is known to harbour tens of thousands of distinctive genes (Gadgil, 1992).

The earth abounds in a particular range of living organisms. They are customarily classified into five "kingdoms", of which the first is Protistra (bacteria and blue-green algae); the second, Monera (advanced algae); third, Fungi; fourth Plants; and fifth, Animals.

Living organisms are found everywhere–from the depths of the sea to tropical rain forests and from the polar icecaps to hot springs. They range in size from viruses that may be no bigger than one-millionth of a metre to the African elephant, which can weigh upto 6.5 tonnes, and a redwood tree which is more than 100 metres tall.

A particularly surprising aspect of biodiversity is its apparent "lopsidedness". An overwhelming proportion of the animal biomass in tropical forests in contributed by insects (Fig. 1.1), among them, social insects such as ants and termites. In a Brazilian tropical rainforest, for example, it has been estimated that the biomass of ants is approximately four times that of all vertebrates (i.e., amphibians, reptiles, birds and mammals) put together. Few realise that in tropical forests, the role of higher animals such as vertebrates is "insignificant", compared to ants and other insects.

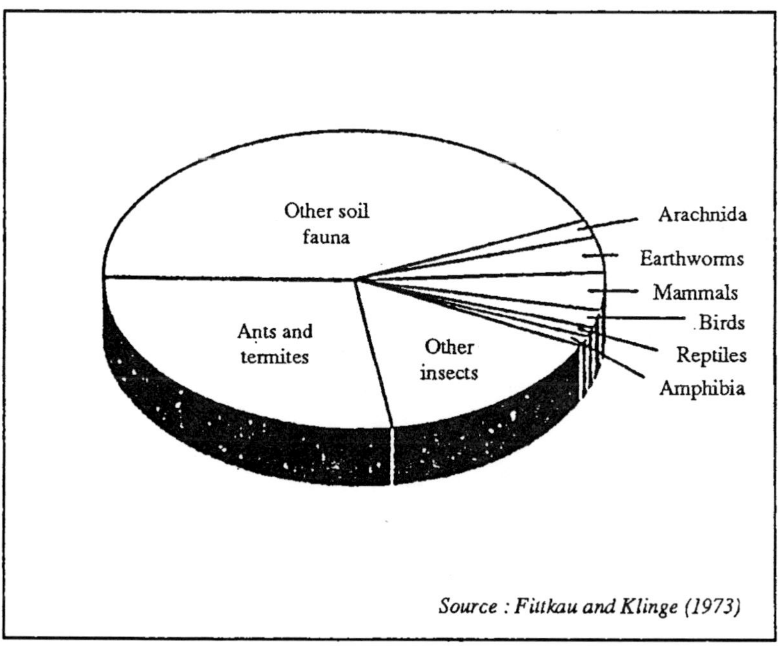

Fig. 1.1. Total animal biomass.

It is not easy to record, describe or even count the number of extent species. Life-forms are so diverse that many years of specialised training are required to be able to recognise and describe them and even then, such training will provide knowledge of just one small group of living organisms. Scientists have been meticulously naming and describing living organisms for about 200 years. Indian taxonomists have so far named and described about 83,000 species (Fig. 1.2), while the total for taxonomists world-wide is about 1.8 million (Fig. 1.3). From these numbers estimates were made setting the total number of species of living organisms in the world at between 5 million and 10 million.

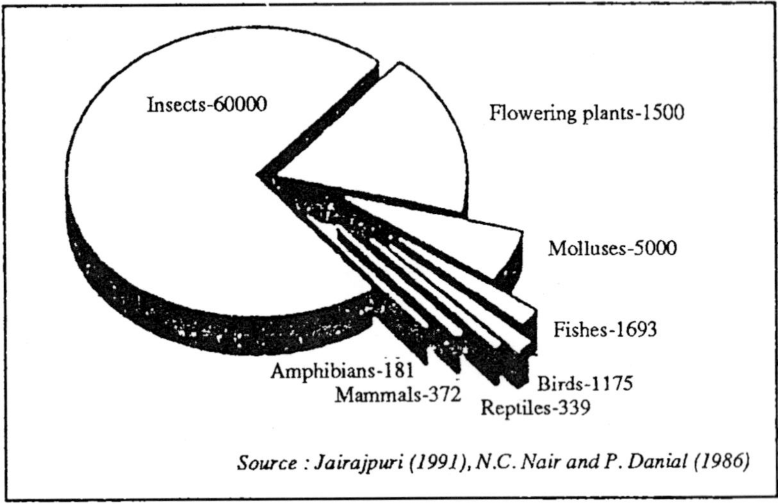

Fig. 1.2. Described Indian species.

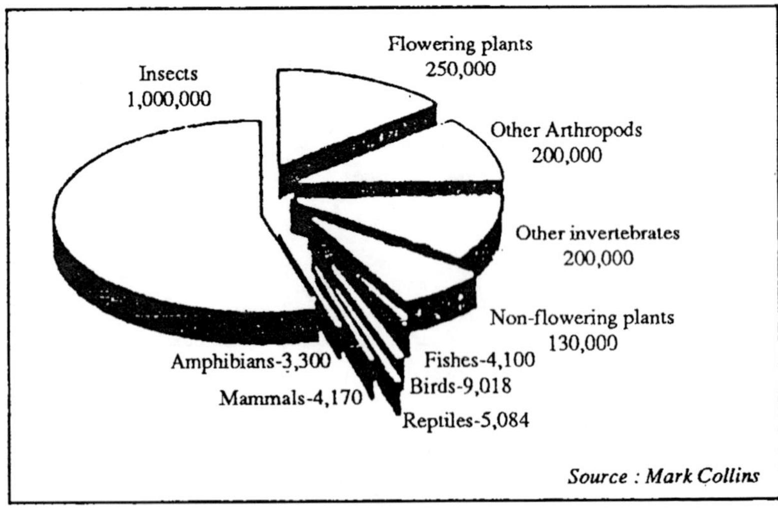

Fig. 1.3. Described World species.

The estimation of biodiversity is perhaps the greatest challenge for it would require millions of man-days of painstaking work by highly trained biologists. More difficult, it would require foresight and imagination on the part of science administrators and the realisation that all this is a worthwhile exercise.

Ours is a tropical country with a tremendous heterogeneity of environments ranging from tropical rain forests of Andaman and Arunachal Pradesh to the hot deserts of Rajasthan and cold desert of Ladakh. It lies at the junction of three biogeographical provinces of Africa, temperate Eurasia and the Orient. As a result, it has a rich biological heritage that qualifies it as one of the 12 megadiversity nations of the world (Gadgil, 1992). We have about 45,584 species of plants and 49,778 species of animals (Table 1.1).

Table 1.1 : India's Biological Wealth

Plants	*Species*	*Animals*	*Species*
Flowering plants	15000	Mammals	372
Pteridophytes	900	Birds	1228
Bryophytes	2584	Reptiles	428
Lichens	1600	Amphibians	204
Fungi	23000	Fishes	2546
Algae	2500	Mollusces	5000
Total	45584	Insects	40000
		Total	49778

BIODIVERSITY DISAPPEARANCE

No one knows for certain how fast genetically distinct populations and species of other organisms are disappearing, but the rates are far too high now and are accelerating. This is clear from two things: First, most organisms are highly adapted to their habitat; if a habitat is changed dramatically, most or all of the plants, animals and micro-organisms that once occupied it will depart of die out. Second, humanity today is on a rampage of changing natural habitats dramatically cutting them down, ploughing them up, overgrazing them, building on them, damming them, spilling oil into them, changing their climates, exposing them to increased ultraviolet radiation, and so on. That the extinction rate is rapidly increasing can be seen simply from statistics on the destruction of tropical forests, the locus of at least half of the planet's biodiversity. The rate of tropical deforestation in 1989 was almost double that in 1979, with roughly 1.8 per cent of the remaining forests disappearing per year (Ehrlich, 1991).

Among the most frightening data pointing to the urgency of dealing with the extinction problem are those on the human impact on global net primary productivity. Net primary production (NPP) is the energy fixed by photosynthesis above and beyond that required to maintain green plants; one can think of NPP as roughly the total food supply of all animals and decomposers.

Since the overwhelming majority (probably over 99%) of the diversity of species now exists on land, the 40 per cent human approximation there alone shows why there is an extinction crises. The amount of energy available for the millions of other kinds of animals clearly has been greatly reduced. Furthermore, the human population is projected to double in the next half-century or so to over 10 billion people.

The implications of the exploding human population is disappearance of the world's biodiversity. Habitat destruction will be greatly accelerated as more and more relatively pristine habitat is converted into urban sprawl, farmland, pastures, reservoirs and like to service the growing human economy.

This sort of direct loss is symbolized by the dramatic decline of forest dwelling songbirds in eastern North America, a consequence of habitat destruction both there and in their Latin American breeding grounds. But even more deadly in the long run will be the habitat degradation likely to result from the combined (and probably synergistic) impact on habitats from rapid climatic change, acid precipitation, increased ultraviolet-beta radiation, and a general spread of toxic substances in the environment.

It remains to be seen what the scale of each of these broad ecosystemic assaults will be. There is debate in the climatological community about the exact rate of global warming and the resultant patterns of climatic change; meanwhile, there has not even been a start at abating emissions of carbon dioxide, nitrous oxide or methane. Since a great deal of inertia is built into the climatic system by the thermal capacity of the oceans, even if steps to slow the greenhouse gas build-up are begun soon, natural communities, especially in temperate and polar regions, have at least even odds of encountering an unprecedented rate of climatic change. And if the warming should trigger more positive then negative feedback mechanisms, as some believe likely, then the rate and degree of change could be catastrophic regardless of human efforts to counter it.

Attempts are being made to deal with various forms of toxic pollution (pesticides, radioactive materials, industrial wastes etc.) but action to abate acid deposition has been slow in coming. It is likely to be difficult, as the human economy expands, to reduce the release of toxic substances. Reductions in emissions per capita will be offset by population growth. Industrialization in poor countries will increase emissions, and highly stressed agricultural systems will require more and more toxic "inputs". The only really cheering sign in the arena of indirect assaults on biodiversity is the recent agreement to halt the production of ozone-destroying chlorofluorocarbons and halons. Even in this case, the level of compliance with the protocol will not be known for some time, and the precise amount of damage entrained by the quantities already released remains to be seen.

But with all uncertainties, it seems safe to assume that the majority of organisms will suffer greatly from the impacts of these interacting factors. The biotas of many lakes already shows serious losses, forests are dying in Europe and North America, and there have been widespread decline in amphibian populations. Climate change alone will cause many organisms simply to fade quitely away, especially since human - disturbed areas will severely limit their possibilities of migrating to keep up with changing climatic conditions (Ehrlich, 1991).

We have often ruthlessly wiped out life in all its diversity. Lord Krishna and Arjuna reduced to ashes the great Khandava forest along with every animal living therein. This is just one example, there are many more to prove that our people have used and abused life with all its diversity over the ages.

The British deliberately destroyed our traditional system of resource use, granting no rights but only some privileges to the local communities over common lands. As a consequence such lands became "no man's lands" abused by all but protected by none. This situation worsened after independence. Yet a great deal of the biodiversity remains protected - thanks to our folk traditions (Gadgil, 1991).

Human beings have the capacity to alter the parts of the habitats with the invention of powerful machinery, large scale burning of fossil fuel combined with large scale production, use and abuse of chemicals. Human beings have greatly increased capacity to alter the rate and extent of habitat change. Number of species of animals and plants have no longer been able to cope with these changes and have become extinct. It has been estimated that 140 plants and animal species are lost every day in the world.

The roots of biodiversity destruction lie not so much in population increase, as in the relations between the communities within each nation, and between the nations themselves. This is responsible for cornering the vast biological resources for the benefit of a small minority within the poor nations, and for the wasteful consumption patterns of the North. Eighteen million hectares of the Amazonian forest has been cleared in Brazil to meet the European and American coffee demand. Germany causes the degradation of 200,000 hectares of rainforest a year for timber. Adverse terms of trade, protectionist policies of the North, dumping of environmentally destructive technologies and materials in the South, and a host of other factors continue to cause severe widespread biodiversity destruction.

The same goes, of course, for the exploitative policies followed by elites within the Southern countries. Vast natural habitats have been plundered to meet the ever growing needs of this minority, aided by laws which legitimise urban-industrial control over resources. The poor are forced to overstrain the meagre resources that are left in their control, and are then portrayed as ecological culprits. In countries like India, the development policies and projects have rarely been sensitive to the needs of the local communities. The government's failure to remove poverty and crub middle-class consumerism has led to conditions in which sensible natural resources management assumes low priority (Kothari, 1993).

The loss of earth's biodiversity is one of the most pressing environmental and developmental issues. With each species that disappears, developing countries loose potential for sustainable development. At the Earth Summit, June 1992, the majority of the world's nations signed a Convention on Biological Diversity. Indeed the convention reflects many concerns, regional and international conservation measures, *in-situ* (on site) and *ex-situ* (off site) protection, role of local communities, transfer of technologies and funds, and aspects of biotechnology and intellectual property rights.

CONVENTION ON BIODIVERSITY

Preamble

The Contracting Parties,

Conscious of the intrinsic value of biological diversity and of the ecological, genetic, social, economic, scientific, educational, cultural recreational and aesthetic values of biological diversity and its components.

Conscious also of the importance of biological diversity for evolution and for maintaining life sustaining systems of the biosphere.

Affirming that the conservation of biological diversity is a common concern of humankind.

Reaffirming that States have sovereign rights over their own biological resources.

Reaffirming also that States are responsible for conserving their biological diversity and for using their biological resources in a sustainable manner.

Concerned that biological diversity is being significantly reduced by certain human activities.

Aware of the general lack of information and knowledge regarding biological diversity and of the urgent need to develop scientific, technical and institutional capacities to provide the basic understanding upon which to plan and implement appropriate measures.

Noting that it is vital to anticipate, prevent and attack the causes of significant reduction of loss of biological diversity at source.

Noting also that where there is a threat of significant reduction or loss of biological diversity, lack of full scientific certainty should not be used as a reason for postponing measures to avoid or minimise such a threat.

Noting further that the fundamental requirement for the conservation of biological diversity is the *in-situ* conservation of ecosystems and natural habitats and the maintenance and recovery of viable populations of species in their natural surroundings.

Noting further that *ex-situ* measures, preferable in the country of origin, also have an important role to play.

Recognising the close and traditional dependence of many indigenous and local communities embodying traditional lifestyles on biological resources, and the desirability of sharing equitably benefits arising from the use of traditional knowledge, innovations and practices relevant to the conservation of biological diversity and the sustainable use of its components.

Recognising also the vital role that women play in the conservation and sustainable use of biological diversity and affirming the need for the full participation of women at all levels of policy-making and implementation for biological diversity conservation.

Stressing the importance of, and the need to promote, international, regional and global cooperation among States and intergovernmental organisations and the non-governmental sector for the conservation of biological diversity and the sustainable use of its components.

Acknowledging that the provision of new and additional financial resources and appropriate access to relevant technologies can be expected to make a substantial difference in the world's ability to address the loss of biological diversity.

Acknowledging further that special provision is required to meet the needs of developing countries, including the provision of new and additional financial resources and appropriate access to relevant technologies.

Noting in this regard the special conditions of the least developed countries and small island States.

Acknowledging that substantial investments are required to conserve biological diversity and there is that expectation of a broad range of environmental, economic and social benefits from those investments.

Recognising that economic and social development and poverty eradication are the first and overriding priorities of developing countries.

Aware that conservation and sustainable use of biological diversity is of critical importance for meeting the food, health and other needs of the growing world population, for which purpose access to and sharing of both genetic resources and technologies are essential.

Noting that, ultimately, the conservation and sustainable use of biological diversity will strengthen friendly relations among States and contribute to peace for humankind.

Desiring to enhance and complement existing international arrangements for the conservation of biological diversity and sustainable use of its components, and

Determined to conserve and sustainably use biological diversity for the benefit of present and future generations,

Have agreed as follows:

Article 1. Objectives

The objectives of this Convention, to be pursued in accordance with its relevant provisions, are the conservation of biological diversity, the sustainable use of its components and the fair and equitable sharing of the benefits arising out of the utilisation of genetic resources, including by appropriate access to genetic resources and by appropriate transfer of relevant technologies, taking into account all rights over those resources and to technologies, and by appropriate funding.

Article 2. Use of Terms

For the purposes of this Convention:

"Biological diversity" means the variability among living organisms from all sources including, inter alia, terrestrial, marine and other aquatic ecosystems and the ecological complexes of which they are part; this includes diversity within species, between species and of ecosystems.

"Biological resources" includes genetic resources organisms or parts thereof, populations, or any other biotic component of ecosystems with actual or potential use or value for humanity.

"Biotechnology" means any technological application that uses biological systems, living organisms, or derivatives thereof, to make or modify products or processes for specific use.

"Country of origin of genetic resources" means the country which possesses those genetic resources in *in-situ* conditions.

"Country providing genetic resources" means the country supplying genetic resources collected from *in-situ* sources, including populations of both wild and domesticated species, or taken from *ex-situ* sources, which may or may not have originated in that country.

"Domesticated or cultivated species" means species in which the evolutionary process has been influenced by humans to meet their needs.

"Ecosystem" means a dynamic complex of plant, animal and micro-organism communities and their non-living environment interacting as a functional unit.

"*Ex-situ* conservation" means the conservation of components of biological diversity outside their natural habitats.

"Genetic material" means any material of plant, animal, microbial or other origin containing functional units of heredity.

"Genetic resources" means genetic material of actual or potential value.

"Habitat" means the place or type of site where an organism or population naturally occurs.

"*In-situ* conditions" means conditions where genetic resources exist within ecosystems and natural habitats, and, in the case of domesticated or cultivated species, in the surroundings where they have developed their distinctive properties.

"*In-situ* conservation" means the conservation of ecosystems and natural habitats and the maintenance and recovery of viable populations of species in their natural surroundings and, in the case of domesticated or cultivated species, in the surroundings where they have developed their distinctive properties.

"Protected area" means a geographically defined area which is designated or regulated and managed to achieve specific conservation objectives.

"Regional economic integration organisation" means an organisation constituted by sovereign States of a given region, to which its member States have transferred competence in respect of matters governed by this Convention and which has been duly authorised, in accordance with its internal procedures, to sign, ratify, accept, approve or accede to it.

"Sustainable use" means the use of components of biological diversity in a way and at a rate that does not lead to the long-term decline of biological diversity, thereby maintaining its potential to meet the needs and aspirations of present and future generations.

"Technology" includes biotechnology.

Article 3. Principle

States have, in accordance with the Charter of the United Nations and the principles of international law, the sovereign right to exploit their own resources pursuant to their own environmental policies, and the responsibility to ensure that activities within their jurisdiction or control do not cause damage to the environment of other States or of areas beyond the limits of national jurisdiction.

Article 4. Jurisdictional Scope

Subject to the rights of other States, and except as otherwise expressly provided in this Convention, the provisions of this Convention apply, in relation to each Contracting Party:

(a) In the case of components of biological diversity, in areas within the limits of its national jurisdiction; and

(b) In the case of processes and activities, regardless of where their effects occur, carried out under its jurisdiction or control within the areas of its national jurisdiction or beyond the limits of national jurisdiction.

Article 5. Cooperation

Each contracting party shall, as far as possible and as appropriate, cooperate with other Contracting Parties, directly or, where appropriate, through competent international organisations, in respect of areas beyond national jurisdiction and on other matters of mutual interest, for the conservation and sustainable use of biological diversity.

Article 6. General Measures for Conservation and Sustainable Use

Each Contracting Party shall, in accordance with its particular conditions and capabilities:

(a) Develop national strategies, plans or programmes for the conservation and sustainable use of biological diversity or adapt for this purpose existing strategies, plans or programmes which shall reflect, *inter alia,* the measures set out in this convention relevant to the Contracting Party concerned; and

(b) Integrate, as far as possible and as appropriate, the conservation and sustainable use of biological diversity into relevant sectoral or cross-sectoral plans, programmes and policies.

Article 7. Identification and Monitoring

Each Contracting Party shall, as far as possible and as appropriate, in particular for the purposes of Articles 8 to 10:

(a) Identify components of biological diversity important for its conservation and sustainable use having regard to the indicative list of categories set down in Annex I;

(b) Monitor, through sampling and other techniques, the components of biological diversity identified pursuant to subparagraph (a) above, paying particular attention to those requiring urgent conservation measures and those which offer the greatest potential for sustainable use;

(c) Identify processes and categories of activities which have or are likely to have significant adverse impacts on the conservation and sustainable use of biological diversity, and monitor their effects through sampling and other techniques; and

(d) Maintain and organise, by any mechanism data, derived from identification and monitoring activities pursuant to subparagraphs (a), (b) and (c) above.

Article 8. *In-situ* Conservation

Each contracting party shall as far as possible and as appropriate:

(a) Establish a system of protected areas or areas where special measures need to be taken to conserve biological diversity;

(b) Develop, where necessary, guidelines for the selection, establishment and management of protected areas or areas where special measures need to be taken to conserve biological diversity.

(c) Regulate or manage biological resources important for the conservation of biological diversity whether within or outside protected areas, with a view to ensuring their conservation and sustainable use;

(d) Promote the protection of ecosystems, natural habitats and the maintenance of viable populations of species in natural surroundings;

(e) Promote environmentally sound and sustainable development in areas adjacent to protected areas with a view to furthering protection of these areas;

(f) Rehabilitate and restore degraded ecosystems and promote the recovery of threatened species, *inter alia*, through the development and implementation of plans or other management strategies;

(g) Establish or maintain means to regulate, manage or control the risk associated with the use and release of living modified organisms resulting from biotechnology which are likely to have adverse environmental impacts that could affect the conservation and sustainable use of biological diversity, taking also into account the risks to human health;

(h) Prevent the introduction of control or eradicate those alien species which threaten ecosystems, habitats or species;

(i) Endeavour to provide the conditions needed for compatibility between present uses and the conservation of biological diversity and the sustainable use of its components;

(j) Subject to its national legislation, respect, preserve and maintain knowledge, innovations and practices of indigenous and local communities embodying traditional lifestyles relevant for the conservation and sustainable use of biological diversity and promote their wider application with the approval and involvement of the holder of such knowledge, innovations and practices and encourage the equitable sharing of the benefits arising from the utilisation of such knowledge, innovations and practices;

(k) Develop or maintain necessary legislation and/or other regulatory provisions for the protection of threatened species and populations;

(l) Where a significant adverse effect on biological diversity has been determined pursuant to Article 7, regulate or manage the relevant processes and categories of activities; and

(m) Cooperate in providing financial and other support for *in-situ* conservation outlined in subparagraphs (a) to (l) above, particularly to developing countries.

Article 9. *Ex-situ* Conservation

Each Contracting Party shall as far as possible and as appropriate, and predominantly for the purpose of complementing *in-situ* measures;

(a) Adopt measures for the *ex-situ* conservation of components of biological diversity, preferably in the country of origin of such components;

(b) Establish and maintain facilities for *ex-situ* conservation of and research on plants, animals and micro-organisms, preferably in the country of origin of genetic resources;

(c) Adopt measures for the recovery and rehabilitation of threatened species and for their reintroduction into their natural habitats under appropriate conditions;

(d) Regulate and manage collection of biological resources from natural habitats for *ex-situ* conservation purposes so as not to threaten ecosystems and *in-situ* populations of species, except where special temporary *ex-situ* measures are required under subparagraph (c) above; and

(e) Cooperate in providing financial and other support for *ex-situ* conservation outlined in subparagraphs (a) to (d) above and in the establishment and maintenance of *ex-situ* conservation facilities in developing countries.

Article 10. Sustainable Use of Components of Biological Diversity

Each contracting party shall, as far as possible and as appropriate;

(a) Integrate consideration of the conservation and sustainable use of biological resources into national decision-making;

(b) Adopt measures relating to the use of biological resources to avoid or minimise adverse impacts on biological diversity;

(c) Protect and encourage customary use of biological resources in accordance with traditional cultural practices that are compatible with conservation or sustainable use requirements;

(d) Support local populations to develop and implement remedial action in degraded areas where biological diversity has been reduced; and

(e) Encourage cooperation between its governmental authorities and its private sector in developing methods for sustainable use of biological resources.

Article 11. Incentive Measures

Each Contracting Party shall as far as possible and as appropriate, adopt economically and socially sound measures that act as incentives for the conservation and sustainable use of components of biological diversity.

Article 12. Research and Training

The Contracting Parties, taking into account the special needs of developing countries, shall:

(a) Establish and maintain programmes for scientific and technical education and training in measures for the identification, conservation and sustainable use of biological diversity and its components and provide support for such education and training for the specific needs of developing countries;

(b) Promote and encourage research which contributes to the conservation and sustainable use of biological diversity, particularly in developing countries, *inter alia,* in accordance with decisions of the Conference of the Parties taken in consequence of recommendations of the subsidiary Body on Scientific, Technical and Technological Advice; and

(c) In keeping with the provisions of Articles 16, 18 and 20, promote and cooperate in the use of scientific advances in biological diversity research in developing methods for conservation and sustainable use of biological resources.

Article 13. Public Education and Awareness

The Contracting Parties shall:

(a) Promote and encourage understanding of the importance of, and the measures required for, the conservation of biological diversity, as well as its propagation through media, and the inclusion of these topics in educational programmes; and

(b) Cooperate, as appropriate, with other States and international organisations in developing educational and public awareness programmes, with respect to conservation and sustainable use of biological diversity.

Article 14. Impact Assessment and Minimising Adverse Impacts

1. Each Contracting Party, as far as possible and as appropriate, shall:

(a) Introduce appropriate procedures requiring environmental impact assessment of its proposed projects that are likely to have significant adverse effects on biological diversity with a view to avoiding or minimising such effects and where appropriate, allow for public participation in such procedures;

(b) Introduce appropriate arrangements to ensure that the environmental consequences of its programmes and policies that are likely to have significant adverse impacts on biological diversity are duly taken into account;

(c) Promote, on the basis of reciprocity, notification, exchange of information and consultation on activities under their jurisdiction or control which are likely to significantly affect adversely the biological diversity of other States or areas beyond the limits of national jurisdiction, by encouraging the conclusion of bilateral, regional or multilateral arrangements, as appropriate;

(d) In the case of imminent or grave danger of damage, origination under its jurisdiction or control, to biological diversity within the area under jurisdiction of other States or in areas beyond the limits of national jurisdiction, notify immediately the potentially affected States of such danger or damage, as well as initiate action to prevent or minimise such danger or damage; and

(e) Promote national arrangements for emergency responses to activities or events, whether caused naturally or otherwise, which present a grave and imminent danger to biological diversity and encourage international cooperation to supplement such national efforts and, where appropriate and agreed by the States or regional economic integration organisations concerned, to establish joint contingency plans.

2. The Conference of the Parties shall examine, on the basis of studies to be carried out, the issue of liability and redress, including restoration and compensation, for damage to biological diversity except where such liability is a purely internal matter.

Article 15. Access to Genetic Resources

1. Recognising the sovereign rights of States over their natural resources, the authority to determine access to genetic resources rests with the national Governments and is subject to national legislation.

2. Each Contracting Party shall endeavour to create conditions to facilitate access to genetic resources for environmentally sound uses by other Contracting Parties and not to impose restrictions that run counter to the objectives of this Convention.

3. For the purpose of this Convention, the genetic resources being provided by a Contracting Party, as referred to in this Article and Articles 16 and 19, are only those that are provided by Contracting Parties that are countries of origin of such resources or by the Parties that have acquired the genetic resources in accordance with this Convention.

4. Access, where granted, shall be on mutually agreed terms and subject to the provisions of this Article.

5. Access to genetic resources shall be subject to prior informed consent of the Contracting Party providing such resources, unless otherwise determined by that Party.

6. Each Contracting Party shall endeavour to develop and carry out scientific research based on genetic resources provided by other Contracting Parties with the full participation of and where possible in, such Contracting Parties.

7. Each Contracting Party shall take legislative, administrative or policy measures, as appropriate, and in accordance with Articles 16 and 19 and where necessary, through the financial mechanism established by Articles 70 and 21 with the aim of sharing in a fair and equitable way and results of research and development and the benefits arising from the commercial and other utilisation of genetic resources with the Contracting Party providing such resources. Such sharing shall be upon mutually agreed terms.

Article 16. Access to and Transfer of Technology

1. Each Contracting Party, recognising that technology includes biotechnology, and that both access to and transfer of technology among Contracting Parties are essential elements for the attainment of the objectives of this Convention undertakes subject to the provisions of this Article to provide and/or facilitate access for and transfer to other Contracting Parties of technologies that are relevant to the conservation and sustainable use of biological diversity or make use of genetic resources and do not cause significant damage to the environment.

2. Access to and transfer of technology referred to in paragraph 1 above to developing countries shall be provided and/or facilitated under fair and most favourable terms, including on concessional and preferential terms where mutually agreed, and, where necessary, in accordance with the financial mechanism established by Articles 20 and 21. In the case of technology subject to patents and other intellectual property rights, such access and transfer shall be provided on terms which recognise and are consistent with the adequate and effective protection of intellectual property rights. The application of this paragraph shall be consistent with paragraphs 3, 4 and 5 below.

3. Each Contracting Party shall take legislative, administrative or policy measures, as appropriate, with the aim that Contracting Parties, in particular those that are developing countries, which provide genetic resources are provided access to and transfer to technology which makes use of those resources, on mutually agreed terms, including technology protected by patents and other intellectual property rights, where necessary, through the provisions of Articles 20 and 21 and in accordance with international law and consistent with paragraphs 4 and 5 below.

4. Each Contracting Party shall take legislative, administrative or policy measures as appropriate, with the aim that the private sector facilitates access to, joint development and transfer of technology referred to in paragraph 1 above for the benefit of both governmental institutions and the private sector of developing countries and in this regard shall abide by the obligations included in paragraphs 1, 2 and 3 above.

5. The Contracting Parties, recognising that patents and other intellectual priority rights may have an influence on the implementation of this Convention, shall cooperate in this regard subject to national legislation and international law in order to ensure that such rights are supportive of and do not run counter to its objectives.

Article 17. Exchange of Information

1. The Contracting Parties shall facilitate the exchange of information from all publicly available sources, relevant to the conservation and sustainable use of biological diversity, taking into account the special needs of developing countries.

2. Such exchange of information shall include exchange of results of technical, scientific and socio-economic research, as well as information on training and surveying programmes, specialised knowledge, indigenous and traditional knowledge as such and in combination with the technologies referred to in Article 16, paragraph 1. It shall also, where feasible, include repatriation of information.

Article 18. Technical and Scientific Cooperation

1. The Contracting Parties shall promote international, technical and scientific cooperation in the field of conservation and sustainable use of biological diversity, where necessary, through the appropriate international and national institutions.

2. Each Contracting Party shall promote technical and scientific cooperation with other Contracting Parties, in particular developing countries, in implementing this Convention, *inter alia,* through the development and implementation of national policies. In promoting such cooperation, special attention should be given to the development and strengthening of national capabilities, by means of human resources development and institution building.

3. The Conference of the Parties, at its first meeting, shall determine how to establish a clearing-house mechanism to promote and facilitate technical and scientific cooperation.

4. The Contracting Parties shall, in accordance with national legislation and policies, encourage and develop methods of cooperation for the development and use of technologies, including indigenous and traditional technologies, in pursuance of the objectives of this Convention. For this purpose, the Contracting Parties shall also promote cooperation in the training of personnel and exchange of experts.

5. The Contracting Parties shall, subject to mutual agreement, promote the establishment of joint research programmes and joint ventures for the development of technologies relevant to the objectives of this Convention.

Article 19. Handling of Biotechnology and Distribution of its Benefits

1. Each Contracting Party shall take legislative, administrative or policy measures, as appropriate, to provide for the effective participation in biotechnological research activities by those Contracting Parties, especially developing countries, which provide the genetic resources for such research, and where feasible in such Contracting Parties.

2. Each Contracting Party shall take all practicable measures to promote and advance priority access on a fair and equitable basis by Contracting Parties, especially developing countries, to the results and benefits arising from biotechnologies based upon genetic resources provided by those Contracting Parties. Such access shall be on mutually agreed terms.

3. The Parties shall consider the need for the modalities of a protocol setting out appropriate procedures, including, in particular, advance informed agreement, in the field of the safe transfer, handing and use of any living modified organism resulting from biotechnology that may have adverse effect on the conservation and sustainable use of biological diversity.

4. Each Contacting Party shall, directly or by requiring any natural or legal person under its jurisdiction providing the organisms referred to in paragraph 3 above, provide any available information about the use and safety regulations required by that Contracting Party in handling such organisms, as well as any available information on the potential adverse impact of the specific organisms concerned to the Contracting Party into which those organisms are to be introduced.

Article 20. Financial Resources

1. Each Contracting Party undertakes to provide, in accordance with its capabilities, financial support and incentives in respect of those national activities which are intended to achieve the objectives of this Convention, in accordance with its national plans, priorities and programmes.

2. The developed country Parties shall provide new and additional financial resources to enable developing country Parties to meet the agreed full incremental costs to them of implementing measures which fulfil the obligations of this Convention and to benefit from its provisions and which costs are agreed between a developing country Party and the institutional structure referred to in Article 21, in accordance with policy, strategy, programme priorities and eligibility criteria and an indicative list of

incremental costs established by the Conference of the Parties. Other Parties, including countries undergoing the process of transition to a market economy, may voluntarily assume the obligations of the developed country Parties. For the purpose of this Article, the Conference of the Parties, shall at its first meeting establish a list of developed country Parties and other Parties which voluntarily assume the obligations of the developed country Parties. The Conference of the Parties shall periodically review and if necessary amend the list Contributions from other countries and sources on a voluntary basis would also be encouraged. The implementation of these commitments shall take into account the need for adequacy, predictability and timely flow of funds and the importance of burden-sharing among the contributing Parties included in the list.

3. The developed country Parties may also provide, and developing country Parties avail themselves of financial resources related to the implementation of this Convention through bilateral, regional and other multilateral channels.

4. The extent to which developing country Parties will effectively implement their commitments under this Convention will depend on the effective implementation by developed country Parties of their commitments under this Convention related to financial resources and transfer of technology and will take fully into account the fact that economic and social development and eradication of poverty are the first and overriding priorities of the developing country Parties.

5. The Parties shall take full account of the specific needs and special situation of least developed countries in their actions with regard to funding the transfer of technology.

6. The Contracting Parties shall also take into consideration the special conditions resulting from the dependence on distribution and location of, biological diversity within developing country Parties, in particular small island States.

7. Consideration shall also be given to the special situation of developing countries, including those that are most environmentally vulnerable, such as those with arid and semi-arid zones, coastal and mountainous areas.

Article 21. Financial Mechanism

1. There shall be a mechanism for the provision of financial resources to developing country Parties for purposes of this Convention on

a grant or concessional basis the essential elements of which are described in this Article. The mechanism shall function under the authority and guidance of, and be countable to, the Conference of the Parties for purposes of this Convention. The operations of the mechanism shall be carried out by such institutional structure as may be decided upon by the Conference of the Parties at its first meeting. For purposes of this Convention, the Conference of the Parties shall determine the policy, strategy, programme priorities and eligibility criteria relating to the access to and utilisation of such resources. The contributions shall be such as to take into account the need for predictability, adequacy and timely flow of funds referred to in Article 20 in accordance with the amount of resources needed to be decided periodically by the Conference of the Parties and the importance of burden-sharing among the contributing Parties included in the list referred to in Article 20, paragraph 2. Voluntary contributions may also be made by the developed country Parties and by other countries and sources. The mechanism shall operate within a democratic and transparent system of governance.

2. Pursuant to the objective of this Convention, the Conference of the Parties shall at is first meeting determine the policy, strategy and programme priorities, as well as detailed criteria and guidelines for eligibility for access to and utilisation of the financial resources including monitoring and evaluation on a regular basis of such utilisation. The Conference of the Parties shall decide on the arrangements to give effect to paragraph 1 above after consultation with the institutional structure entrusted with the operation of the financial mechanism.

3. The Conference of the Parties shall review the effectiveness of the mechanism established under this Article, including the criteria and guidelines referred to in paragraph 2 above, not less than two years after the entry into force of this Convention and thereafter on a regular basis. Based on such review, it shall take appropriate action to improve the effectiveness of the mechanism if necessary.

4. The Contracting Parties shall consider strengthening existing financial institutions to provide financial resources for the conservation and sustainable use of biological diversity.

Article 22. Relationship with Other International Conventions

1. The provisions of this Convention shall not affect the rights and obligations of any Contracting Party deriving from any existing international agreement, except where the exercise of those rights and obligations would cause a serious damage or threat to biological diversity.

2. Contracting Parties shall implement this Convention with respect to the marine environment consistently with the rights and obligations of States under the law of the sea.

Article 23. Conference of the Parties

1. A Conference of the Parties is hereby established. The first meeting of the Conference of the Parties shall be convened by the Executive Director of the United Nations Environment Programme not later than one year after the entry into force of this Convention. Thereafter, ordinary meetings of the Conference of the Parties shall be held at regular intervals to be determined by the Conference at its first meeting.

2. Extraordinary meetings of the Conference of the Parties shall be held at such other times as may be deemed necessary by the Conference, or at the written request of any Party, provided that, within six months of the request being communicated to them by the Secretariat, it is supported by at least one third of the Parties.

3. The Conference of the Parties shall be consensus agree upon and adopt rules of procedure for itself and for any subsidiary body it may establish, as well as financial rules governing the funding of the Secretariat. At each ordinary meeting, it shall adopt a budget for the financial period until the next ordinary meeting.

4. The Conference of the Parties shall keep under review the implementation of this Convention, and, for this purpose, shall:

(a) Establish the form and the intervals for transmitting the information to be submitted in accordance with Article 26 and consider such information as well as reports submitted by any subsidiary body;

(b) Review scientific, technical and technological advice on biological diversity provided in accordance with Article 25;

(c) Consider and adopt, as required, protocols in accordance with Article 28;

(d) Consider and adopt, as required, in accordance with Article 29 and 30, amendments to this Convention and its annexes;

(e) Consider amendments to any protocol, as well as to any annexes thereto, and if so decided, recommend their adoption to the parties to the protocol concerned;

(f) Consider and adopt, as required, in accordance with Article 30, additional annexes to this Convention;

(g) Establish such subsidiary bodies, particularly to provide scientific and technical advice, as are deemed necessary for the implementation of this Convention.

(h) Contact, through the Secretariat, the executive bodies of conventions dealing with matters convened by this Convention with a view to establishing appropriate forms of cooperation with them; and

(i) Consider and undertake and additional action that may be required for the achievement of the purposes of this Convention in the light of experience gained in its operation.

5. The United Nations, its specialised agencies and the International Atomic Agency, as well as any State not Party to this Convention, may be represented as observers at meetings of the Conference of the Parties. Any other body or agency, whether governmental or non-governmental, qualified in fields relating to conservation and sustainable use of biological diversity, which has informed the Secretariat of its wish to be represented as an observer at a meeting of the Conference of the Parties, may be admitted unless at least one third of the Parties present object. The admission and participation of observers shall be subject to the rules of procedure adopted by the Conference of the Parties.

Article 24. Secretariat

1. A Secretariat is hereby established. In functions shall be:

(a) To arrange for and service meetings of the Conference of the Parties provided for in Article 23;

(b) To perform the functions assigned to it by any protocol;

(c) To prepare reports on the execution of its functions under this Convention and present them to the Conference of the Parties.

(d) To coordinate with other relevant international bodies and, in particular to enter into such administrative and contractual arrangements as may be required for the effective dischange of its functions; and

(e) To perform such other functions as may be determined by the Conference of the Parties.

2. At its first ordinary meeting, the Conference of the Parties shall designate the secretariat from amongst those existing competent international organisations which have signified their willingness to carry out the secretariat functions under this Convention.

Article 25. Subsidiary Body on Scientific, Technical and Technological Advice

1. A subsidiary body for the provision of scientific, technical and technological advice is hereby established to provide the Conference of the Parties and, as appropriate, its other subsidiary bodies with timely advice relating to the implementation of this Convention. This body shall be open to participation by all Parties and shall be multidisciplinary. It shall comprise government representatives competent in the relevant field of expertise. It shall report regularly to the Conference of the Parties on all aspects of its work.

2. Under the authority of and in accordance with guidelines laid down by the Conference of the Parties, and upon its request, this body shall:

(a) Provide scientific and technical assessments of the status of biological diversity;

(b) Prepare scientific and technical assessment of the effects of types of measures taken in accordance with the provisions of this Convention;

(c) Identify innovative, efficient, and state-of-the-art technologies and know-how relating to the conservation and sustainable use of biological diversity and advise on the ways and means of promoting development and/or transferring such technologies;

(d) Provide advice on scientific programmes and international cooperation in research and development related to conservation and sustainable use of biological diversity; and

(e) Respond to scientific, technical, technological and methodological questions that the Conference of the Parties and its subsidiary bodies may put to the body.

3. The functions, terms of reference, organisation and operation of this body may be further elaborated by the Conference of the Parties.

Article 26. Reports

Each Contracting Party shall, at intervals to be determined by the Conference of the Parties, present to the Conference of the Parties, reports on measures which it has taken for the implementation of the provisions of this Convention and their effectiveness in meeting the objectives of this Convention.

Article 27. Settlement of Disputes

1. In the event of a dispute between Contracting Parties concerning the interpretation or application of this Convention, the parties concerned shall seek solution by negotiation.

2. If the parties concerned cannot reach agreement by negotiation, they may jointly seek the good offices of, or request mediation by, a third party.

3. When ratifying, accepting, approving or acceding to this Convention, or at any time thereafter, a State or regional economic integration organisation may declare in writing to the Depositary that for a dispute not resolved in accordance with paragraph 1 or paragraph 2 above, it accepts one or both of the following means of dispute settlement as compulsory:

(a) Arbitration in accordance with the procedure laid down in Part 1 of Annex II;

(b) Submission of the dispute to the International Court of Justice.

4. If the parties to the dispute have not, in accordance with paragraph 3 above, accepted the same or any procedure, the dispute shall be submitted to conciliation in accordance with Part 2 of Annex II unless the parties otherwise agree.

5. The provisions of this Article shall apply with respect to any protocol except as otherwise provided in the protocol concerned.

Article 28. Adoption of Protocols

1. The Contracting Parties shall cooperate in the formulation and adoption of protocols in this Convention.

2. Protocols shall be adopted at a meeting of the Conference of the Parties.

3. The text of any proposed protocol shall be communicated to the Contracting Parties by the Secretariat at least six months before such a meeting.

Article 29. Amendment of the Convention or Protocols

1. Amendments to this Convention may be proposed by any Contracting Party. Amendments to any protocol may be proposed by any Party to that protocol.

2. Amendments to this Convention shall be adopted at a meeting of the Conference of the Parties. Amendments to any protocol shall be adopted at a meeting of the Parties to the Protocol in question. The text of any proposed amendment to this Convention or to any protocol, except as may otherwise be provided in such protocol, shall be communicated to the Parties to the instrument in question by the secretariat at least six months before the meeting at which it is proposed for adoption. The secretariat shall also communicate proposed amendments to the signatories to this Convention for information.

3. The Parties shall make every effort to reach agreement on any proposed amendment to this Convention or to any protocol by consensus. If all efforts at consensus have been exhausted, and no agreement reached, the amendment shall as a last resort be adopted by a two-third majority vote of the parties to the instrument in question present and voting at the meeting and shall be submitted by the Depositary to all Parties for ratification, acceptance or approval.

4. Ratification, acceptance or approval of amendments shall be notified to the Depositary in writing. Amendments adopted in accordance with paragraph 3 above shall enter into force among Parties having accepted them on the nineteenth day after the deposit of instruments of ratification, acceptance or approval by at least two thirds of the Contracting Parties to this Convention or of the Parties to the protocol concerned, except as may otherwise be provided in such protocol. Thereafter the amendments shall enter into force for any other Party on the ninetieth day after that Party deposits its instrument of ratification, acceptance or approval of the amendments.

5. For the purposes of this Article, "Parties present and voting" means Parties present and casting an affirmative or negative vote.

Article 30. Adoption and Amendment of Annexes

1. The annexes to this Convention or to any protocol shall form an integral part of the Convention or of such protocol, as the case may be, and, unless expressly provided otherwise, a reference to this Convention or its protocols constitutes at the same time a reference to any annexes thereto. Such annexes shall be restricted to procedural, scientific, technical and administrative matters.

2. Except as may be otherwise provided in any protocol with respect to its annexes, the following procedure shall apply to the proposal,

adoption and entry into force of additional annexes to this Convention or of annexes to any protocol:

(a) Annexes to this Convention or to any protocol shall be proposed and adopted according to the procedure laid down in Article 29;

(b) Any party that is unable to approve an additional annex to this Convention or an annex to any protocol to which it is party shall so notify the Depositary, in writing, within one year from the date of the communication of the adoption by the Depositary. The Depositary shall without delay notify all parties of any such notification received. A party may at any time withdraw a previous declaration of objection and the annexes shall thereupon enter into force for that party subject to subparagraph (c) below.

(c) On the expiry of one year from the date of the communication of the adoption by the Depositary, the annex shall enter into force for all parties to this convention or to any protocol concerned which have not submitted a notification in accordance with the provisions of subparagraph (b) above.

3. The proposal, adoption and entry into force of amendments to annexes to this convention or to any protocol shall be subject to the same procedure as for the proposal, adoption and entry into force of annexes to the convention or annexes to any protocol.

4. If an additional annex or an amendment to an annex is related to an amendment to this convention or to any protocol, the additional annex or amendment shall not enter into force until such time as the amendment to the convention or to the protocol concerned enters into force.

Article 31. Right to Vote

1. Except as provided for in paragraph 2 below, each Contracting Party to this Convention or to any protocol shall have one vote.

2. Regional economic integration organisations, in matters within their competence, shall exercise their right to vote with a number of votes equal to the number of their member States which are Contracting Parties to this convention or the relevant protocol. Such organisations shall not exercise their right to vote if their member States exercise theirs, and *vice versa*.

Article 32. Relationship Between This Convention and Its Protocols

1. A State or a regional economic integration organisation may not become a party to a protocol unless it is, or becomes at the same time, a Contracting Party to this Convention.

2. Decisions under any protocol shall be taken only be the parties to the protocol concerned. Any Contracting Party that has not ratified, accepted or approved a protocol may participate as an observer in any meeting of the parties to that protocol.

Article 33. Signature

This Convention shall be open for signature at Rio De Janeiro by all States and any regional economic integration organisation from 5 June, 1992 until 14 June, 1992, and at the United Nations Headquarters in New York from 15 June, 1992 to 4 June, 1993.

Article 34. Ratification, Acceptance or Approval

1. This Convention and any protocol shall be subject to ratification, acceptance or approval by States and by regional economic integration organisations. Instruments of ratification, acceptance or approval shall be deposited with the Depositary.

2. Any organisation referred to in paragraph 1 above which becomes a Contracting Party to this convention or any protocol without any of its member States being a contracting party shall be bound by all the obligations under the convention or the protocol, as the case may be. In the case of such organisations one or more of whose member States is a contracting party to this convention or relevant protocol, the organisation and its member States shall decide on their respective responsibilities for the performance of their obligations under the convention or protocol, as the case may be. In such cases, the organisation and the member States shall not be entitled to exercise rights under the convention or relevant protocol concurrently.

3. In their instruments of ratification, acceptance or approval, the organisations referred to in paragraph 1 above shall declare the extent of their competence with respect to the matters governed by the convention or the relevant protocol. These organisations shall also inform the Depositary of any relevant modification in the extent of their competence.

Article 35. Accession

1. This Convention and any protocol shall be open for accession by States and by regional economic integration organisations from the date on which the convention or the protocol concerned is closed for signature. The instruments of accession shall be deposited with the Depositary.

2. In their instruments of accession, the organisations referred to in paragraph 1 above shall declare the extent of their competence with respect to the matters governed by the Convention or the relevant protocol. These organisations shall also inform the depositary of any relevant modification in the extent of their competence.

3. The provisions of Article 34, paragraph 2, shall apply to regional economic integration organisations which accede to this Convention or any protocol.

Article 36. Entry Into Force

1. This Convention shall enter into force on the ninetieth day after the date of deposit of the thirtieth instrument of ratification, acceptance, approval or accession.

2. Any protocol shall enter into force on the ninetieth day after the date of deposit of the number of instruments of ratification, acceptance, approval or accession, specified in that protocol, has been deposited.

3. For each Contracting Party which ratifies, accepts or approves this convention or accedes thereto after the deposit of the thirtieth instrument of ratification, acceptance, approval or accession, it shall enter into force on the ninetieth day after the date of deposit by such contracting party of its instrument of ratification, acceptance, approval or accession.

4. Any protocol, except as otherwise provided in such protocol, shall enter into force for a contracting party that ratifies, accepts or approves that protocol or accedes thereto after its entry into force pursuant to paragraph 2 above, on the ninetieth day after the date on which that contracting party deposits its instrument of ratification, acceptance, approval or accession, or on the date on which this convention enters into force for that Contracting Party, whichever shall be the later..

5. For the purposes of paragraphs 1 and 2 above, any instrument deposited by a regional economic integration organisation shall not be counted as additional to those deposited by member States of such organisation.

Article 37. Reservations

No reservations may be made to this Convention.

Article 38. Withdrawals

1. At any time after two years from the date on which this Convention has entered into force for a Contracting Party, that contracting party may withdraw from the Convention by giving written notification to the Depositary.

2. Any such withdrawal shall take place upon expiry of one year after the date of its receipt by the Depositary, or on such later date as may be specified in the notification of the withdrawal.

3. Any Contracting Party which withdraws from this convention shall be considered as also having withdrawn from any protocol to which it is party.

Article 39. Financial Interim Arrangements

Provided that it has been fully restructured in accordance with the requirements of Article 21, the Global Environment Facility of the United Nations Development Programme, the United Nations Environment Programme and the International Bank for the Reconstruction and Development shall be the institutional structure referred to in Article 21 on an interim basis, or the period between the entry into force of this convention and the first meeting of the conference of the parties or until the conference of the parties decides which institutional structure will be designated in accordance with Article 21.

Article 40. Secretariat Interim Arrangements

The secretariat to be provided by the Executive Director of the United Nations Environment Programme shall be the secretariat referred to in Article 24, paragraph 2, on an interim basis for the period between the entry into force of this convention and the first meeting of the Conference of the Parties.

Article 41. Depositary

The Secretary-General of the United Nations shall assume the functions of Depositary of this Convention and any protocols.

Article 42. Authentic Texts

The original of this Convention of which the Arabic, Chinese, English, French, Russian and Spanish texts are equally authentic, shall be deposited with the Secretary-General of the United Nations.

In witness whereof the undersigned, being duly authorised to that effect, have signed this Convention.

Done at Rio de Janeiro on this fifth day of June, one thousand nine hundred and ninety two.

ANNEXE I

Identification and Monitoring

1. Ecosystems and habitats; containing high diversity, large numbers of endemic or threatened species, or wilderness; required by migratory species; of social, economic, cultural or scientific importance; or, which are representative, unique or associated with key evolutionary or other biological processes;

2. Species and communities which are: threatened; wild relatives of domesticated or cultivated species; of medicinal, agricultural or other economic value; or social, scientific or cultural importance; or importance for research into the conservation and sustainable use of biological diversity, such as indicator species; and

3. Described genomes and genes of social, scientific or economic importance.

ANNEX II–PART I

Arbitration–Article 1

The claimant party shall notify the secretariat that the parties are referring a dispute to arbitration pursuant to Article 27. The notification shall state the subject-matter of arbitration and include, in particular, the articles of the Convention or the protocol, the interpretation or application of which are at issue. If the parties do not agree on the subject matter of the dispute before the President of the tribunal is designated, the arbitral tribunal shall determine the subject matter. The secretariat shall forward the information thus received to all Contracting Parties to this Convention or to the protocol concerned.

Article 2

1. In disputes between two parties, the arbitral tribunal shall consist of three members. Each of the parties to the dispute shall appoint an arbitrator and the two arbitrators so appointed shall designate by common agreement the third arbitrator who shall be the President of the tribunal. The latter shall not be a national of one of the parties to the dispute, nor have his or her usual place of residence in the territory of one of these parties, nor be employed by any of them, nor have dealt with the case in any other capacity.

2. In disputes between more than two parties, parties in the same interest shall appoint one arbitrator jointly by agreement.

3. Any vacancy shall be filled in the manner prescribed for the initial appointment.

Article 3

1. If the President of the arbitral tribunal has not been designated within two months of the appointment of the second arbitrator, the Secretary-General of the United Nations shall, at the request of a party, designate the President within a further two-month period.

2. If one of the parties to the dispute does not appoint an arbitrator within two months of receipt of the request, the other party may inform the Secretary-General who shall make the designation within a further two-month period.

Article 4

The arbitral tribunal shall render its decisions in accordance with the provisions of this Convention, any protocols concerned, and international law.

Article 5

Unless the parties to the dispute otherwise agree, the arbitral tribunal shall determine its own rules of procedure.

Article 6

The arbitral tribunal may, at the request of one of the parties, recommend essential interim measures of protection.

Article 7

The parties to the dispute shall facilitate the work of the arbitral tribunal and, in particular, using all means at their disposal, shall;

(a) Provide it with all relevant documents, information and facilities, and

(b) Enable it, when necessary, to call witnesses or experts and receive their evidence.

Article 8

The parties and the arbitrators are under an obligation to protect the confidentiality of any information they receive in confidence during the proceedings of the arbitral tribunal.

Article 9

Unless the arbitral tribunal determines otherwise because of the particular circumstances of the case, the costs of the tribunal shall be borne by the parties to the dispute in equal shares. The tribunal shall keep a record of all its costs, and shall furnish a final statement thereof to the parties.

Article 10

Any Contracting Party that has an interest of a legal nature in the subject-matter of the dispute which may be affected by the decision in the case, may intervene in the proceedings with the consent of the tribunal.

Article 11

The tribunal may hear and determine counterclaims arising directly out of the subject-matter of the dispute.

Article 12

Decisions both on procedure and substance of the arbitral tribunal shall be taken by a majority vote of its members.

Article 13

If one of the parties to the dispute does not appear before the arbitral tribunal or fails to defend its case, the other party may request the tribunal to continue the proceedings and to make its award. Absence of a party or a failure of a party to defend its case shall not constitute a bar to the proceedings. Before rendering its final decisions, the arbitral tribunal must satisfy itself that the claim is well founded in fact and law.

Article 14

The tribunal shall render its final decision within five months of the date on which it is fully constituted unless it finds it necessary to extend the time-limit for a period which should not exceed five more months.

Article 15

The final decision of the arbitral tribunal shall be confined to the subject-matter of the dispute and shall state the reasons on which it is based. It shall contain the names of the members who have participated and the date of the final decision. Any member of the tribunal may attach a separate or dissenting opinion to the final decision.

Article 16

The award shall be binding on the parties to the dispute. It shall be without appeal unless the parties to the dispute have agreed in advance to an appellate procedure.

Article 17

Any controversy which may arise between the parties in the dispute as regards the interpretation or manner of implementation of the final decision may be submitted by either party for decision to the arbitral tribunal which rendered it.

PART II - CONCILIATION

Article 1

A conciliation commission shall be created upon the request of one of the parties to the dispute. The commission shall, unless the parties otherwise agree, be composed of five members, two appointed by each Party concerned and a President chosen jointly by those members.

Article 2

In disputes between more than two parties, parties in the same interest shall appoint their members of the commission jointly by agreement. Where two or more parties have separate interests or there is a disagreement as to whether they are of the same interest, they shall appoint their members separately.

Article 3

If any appointments by the parties are not made within two months of the date of the request to create a conciliation commission, the Secretary-General of the United Nations shall, if asked to do so by the party that made the request, make those appointments within a further two-month period.

Article 4

If a President of the conciliation commission has not been chosen within two months of the last of the members of the commission being appointed, the Secretary-General of the United Nations shall, if asked to do so by a party, designate a President within a further two-month period.

Article 5

The conciliation commission shall take its decisions by majority vote of its members. It shall, unless the parties to the dispute otherwise agree, determine its own procedure. It shall render a proposal for resolution of the dispute, which the parties shall consider in good faith.

Article 6

A disagreement as to whether the conciliation commission has competence shall be decided by the commission.

MAJOR ISSUES RELATED TO BIODIVERSITY CONVENTIONS

The government of India, along with 155 other nations signed the Convension on biodiversity at the UNCED Earth Summit held at Rio-de-janerio, Brazil, in June 1992. The Convension deals with various aspects concerning with conservation and sustainable use of all components of biodiversity and affirms that States have sovereign rights over their biological resources.

The agreed text of Convension on biodiversity conceded that all countries have a sovereign right over all biodiversity also had an obligation to take measures for its protection and conservation. It also provided that the developed countries shall have an obligation to provide new and additional funds to developing countries to enable them to carry out the conservation plans. But the provision of access to genetic resources generated much controversy. All countries including developed world except United States of America signed the Convension.

The important issues which need consideration in the context of the Convention for India are :

- Free exchange of germplasm.
- Sustainable use of genetic resources.

FREE EXCHANGE OF GERMPLASM

At present, access is provided through instruments such as the Export-Import Policy, FAO undertaking for crop plant genetic resources and regulations for export and import of seeds. The wild diversity is somehow poorly taken care of. Enforcement of the relevant customs and quarantine officials. As a result, there is unregulated and unscientific erosion of genetic resources. Above all, we do not have any National Register maintaining the details of the genetic resources explored for commercial and research purposes.

In the light of the provisions of the biodiversity convension, it is necessary to consider the following issues :

(a) Creation of a National Register for maintaining the details of the genetic resources exported for commercial or research purposes.

(b) Preparation of model agreements/protocols laying down detailed modalities and conditionalities.

(c) Development of new regulatory mechanisms for providing access to genetic resources.

(d) Regulations for taking out of the country and biomaterial as in Australia.

(e) Mechanism to ensure that the privately held genetic resources are not passed off as new ones under the convension.

(f) Legal declaration from all holders indicating what is lodged at present in their gene banks.

(g) Provision for annuity payments in case of permanent damage to gene banks instead of one time compensation.

(h) Provision for creating State right over genetic resources held privately while framing any new legislation.

(i) Critical examination of the issues related to international Property Rights and patent regimes.

(j) Establishment of an organic linkage between the various agencies involved in activities related to conservation and sustainable use of biodiversity.

INTELLECTUAL PROPERTY RIGHTS

In the package proposed by Arthur Dunkel, the industrialized countries of the North have claimed the right to patent even plant varieties. This innocuous demand contained in Article 3 (b) of Section 5, dealing with patents, if accepted, have devastating consequences for India and other developing countries. If such patenting of plant species and genes is allowed, the ownership of our invaluable genetic resources will pass into the hands of patent holders, mostly multinational companies in the North.

Through this draft developed countries are attempting not only to gain access to these genetic resources but also to establish monopoly control over them through patents. According to the draft, if a new variety is created by a biotechnologist working in a multinational company by transfer of gene "A" for disease resistance from a wheat variety with low yield into another variety "B" with high yield and susceptible to disease, then the multinational company should be able to patent the new variety "C". This means that the new wheat variety "C" will become the sole and exclusive property of multinational company. Not only that, the next generation of the patented variety i.e., the seed is also subject to patent protection. So if there is a high yielding variety of wheat bred by some one in Canada, and a scientist in IARI, New Delhi, India wants to cross it with an Indian disease resistant wheat to improve its yield, he can not do so without paying hefty royalty fees. Even after doing that the new variety automatically becomes the property of the one in Canada.

Until now, genetic resources which are concentrated in tropical and subtropical areas, that is within the sovereign territories of the developing countries, were conveniently regarded as the common heritage of mankind and therefore available to every one free. Patenting of genes gives monopoly rights on life-forms to those who manipulate genes, totally disregarding the contribution of Third World peasants, tribals and healers, for establishing profitable agriculture in their own gene-starved countries without paying for it.

The Northern countries are using trade as a means of enforcing their patent laws and intellectual property rights on the sovereign nations of the Third World. The United States has accused the countries of the Third World of engaging in 'unfair trading practices' if they fail to adopt United States patent laws. Yet it is the United States which has engaged in 'unfair practices' over the use of Third World to turn it into millions of dollars of profits, non of which has been shared with Third World countries.

United States was afraid of the fact that if it signed the biodiversity convension, it would have to share the fruits of researches into biotechnology with the Third World nations in exchange for the wild genes from their tropical forests. They call it as "Intellectual Property Right". But it is not equally unfair and morally unjustified is a developing country from which valuable "genes" of wild seed strains are taken and developed into a "high yielding superior crop" and then asked to buy the same improved seeds of superior crop from the developed nations at a high cost.

A single wild tomato variety (*Lycopersicon chomrelew* Skii) taken from Peru in 1962, has contributed US $ 8 Million a year to the American tomato processing industry by increasing the content of soluble solids in the US tomatoes.

US bred a new high yielding variety of maize (*Zea diploperennis*) worth millions of dollars in value with the help of valuable gene from wild maize species obtained from tropical forests of South. The wild genes also provide ability to adapt against disease, pollution and environmental hazards.

One gene from a single wild barley plant of tropical forest protects US $ 160 million annual barley crops from 'Yellow dwarf virus'.

The Madagascar periwinkle is the source of atleast 60 alkaloids which can treat childhood leukaemia and Hedgkin's disease. Drugs derived from this plant bring about US $ 160 million worth of sales each year.

At the present rate of destruction of the tropical forests, 20-25 percent of the world's plant species will be lost by the year 2000. Consequently, major pharmaceutical companies are now screening and collecting natural plants through contracted third parties. The British Company, Biotics, for example, is a commercial broker known for supplying exotic plants for pharmaceutical screening and inadequately compensating the Third World countries of origin. The Company's officials have actually admitted that many drug companies prefer "sneaking plants" out of the Third World rather than going through legitimate negotiating channels. The US National Cancer Institute has sponsored the World's single largest tropical plant collecting effort by recruiting ethanobotanists to document the traditional medicinal use of plants and other species, yet the indigenous people who willingly share this knowledge are unlikely ever to share in the profit from the development of new drugs or other products (Shiva, 1992).

Intellectual property rights (IPRs) to biological knowledge of northern corporations which are using Third World biodiversity and Third World people's agricultural and medicinal knowledge of plant properties for their manufacture of 'proprietary' products like bio-pesticides and pharmaceuticals. Multinational Companies are expropriating traditional knowledge by patenting of man-based products which are used in India for centuries, since both the tree and knowledge of its pesticides properties have originated in India. Biological resources like them and knowledge of its utilization in health care and agriculture are the collective heritage of Indian farmers and healers, built up over centuries of collective innovation. Over the past few years, American and Japanese companies have taken several patents on formulae for stable neem-based solutions, and even a neem-based toothpaste. Many scientists and ecologists in India regard the American patents on extracts of neem oil as intellectual piracy, for neem-based products, including pesticides, medicines, toothpaste and cosmetics are already in the Indian market, many of them produced in the small scale sector under the Khadi and Village Industries Commission. The assertion of collective IPR made by Indian farmers also pose challenges to the IPR frameworks pushed by the North in multinational platforms such as the General Agreement on Tariffs and Trade (GATT), and the Trade-Related Intellectual Property Rights, the United States interpretation of the biodiversity convension as well as in unilateral threats of the Special 301 clauses of the United States Trade Act.

Continuing apprehension in India

The spectre of "Dunkel" continues to haunt the Indian scene, although his successor, Mr Peter Sutherland, has been in office for over eight months. The latter's entry into the Indian debate on the Uruguay Round Agreement ("Seeds of Doubt", March 15) may spare his predecessor the dubious distinction of being the symbol of all that is undesirable in those agreements. But his defence of the TRIPs (Trade Related Intellectual Property) agreements notice for more substantive reasons.

It is simplistic to suggest that the apprehension in India is based on a misunderstanding "that UPOV is a sui-generis system and the word 'effective' implies something additional". It is misunderstanding to argue that any sui-generis system "allows greater flexibility to adapt the form of protection to particular circumstances". That a suigeneris system means a unique system, not classifiable with others' is known even to the opponents of TRIPs agreement in India. Indeed, the Indian Patents Act, 1970 is a sui-generis system is as much as it leaves out the plant varieties from the scope of patentability altogether. But it is this very uniqueness providing complete freedom to farmers and researchers which will have to be given up because the TRIPs agreement compels the member countries to "provide for the protection of plant varieties either by patents or by an effective sui-generis system or by any combination thereof. Note that sui-generis system is visualised only as an option to the patent system. Moreover, the option must provide "effective" protection. Obviously the "effectiveness" of the system will be judged multilaterally by a mechanism to be established under the new agreement to oversee. the implementation. If this were not so, there would be no need to mention anything at all about the sui-generis system in the multilateral agreement.

No room for doubt

To ensure that there will be no room for doubt in this regard, Article 27.3(b) which provides for the protection for plant varieties adds that "the provision of the subparagraph shall be reviewed four years after the entry into force of the agreement". Clearly the intention is to monitor the "effectiveness" of the implementation. Article 71 further provides that the implementation shall be reviewed by the Council for TRIPs after the expiration of the four-year period of transition. Thereafter such reviews shall take place every two years.

The direction of the process of reviews and the meaning of effectiveness clearly signify progressive upgradation of the levels of protection of plant varieties. But Mr. Sutherland's presentation is conspicuously silent on this aspect. Similarly, he elaborates at length the implications of UPOV 1978 (International Union for Protection of Plant Varieties). But he is economical when he refers to UPOV 1991. He admits that the latter "embodies significantly higher standards of protection" but prefers not to elaborate on the specious ground that it has yet to fulfil its requirement to come into force.

Although Mr. Sutherland is at pains to remove the misgiving that 'sui-generis system' is not UPOV, he does refer to UPOV as laying down the "minimum standards of protection that national system should accord". He further says that "in practice many countries introducing plant variety protection will nontheless wish to profit from the experience that has been developed under UPOV and to tailor their system along the same lines". But when it comes to elaboration, he chooses to slick to the version of UPOV 1978.

New varieties

He underlines that the protection will apply to "new" varieties. It is axiomatic that the variety in public domain cannot claim protection and this not a special virtue of UPOV 1978. In the same vein he says that no protection will be granted unless the plant breeder claims it. But the whole question arose precisely because some influential transnational corporations specialising in plant breeding were demanding the institution of their monopoly rights all over the world. Surely Mr. Sutherland does not think that these corporations would one day turn philanthropic and cease to seek protection in other countries.

The hallowed practice of farmers saving, using and exchanging seeds is referred to somewhat condescendingly as the "so-called" formers privilege. He assures us that the right to save seed for own use will be "allowed". In a country where, not too long ago, thousand of farmers chose to die of starvation but refused to touch the seeds saved for the future crop, this price of saving seeds is not a "so-called" privilege, which it may appear to a lawyer. It is a symbol or norms and values resolutely developed and scrupulously observed by millions of farmers over hundreds of years. For this they do not need grant of permission or recognition by intellectual property lawyers or agreements. It is their way of life.

While recognising this "so-called" privilege, Mr. Sutherland is quick to add that UPOV 1978 is obliged to provide rights in respect of protected seeds that have been produced for commercial marketing. Yes across the fence sakes will be connived at, not so much because it is the traditional practice to be respected as because it will be difficult, if not impossible, to control such activity. But even here, let farmers note: they can continue to indulge in such practice "provided that such traditional practices are not extended to the point where they undermine the protection that should apply to commercial marketing transactions". Clearly the traditional rights are all right so long as they do not threaten to limit or reduce the monopoly profits of the plant breeders. What is paramount is the profit opportunities of transnational corporations. Let it be recalled that according to available estimates, about 60 percent of the seed exchange in the country is carried out by the informal sector of farmers and small operators. Under UPOV 1978 dispensation, such exchanges will be possible for new seeds only with the permission of plant breeders.

This is the situation with UPOV 1978, a rather "gentle" system of the yore. But the world has "progressed" meanwhile. UPOV 1991 is much more demanding. While the scope of UPOV 1978 was limited to 24 plant genera and species to be covered over a period of eight years, UPOV 1991 requires all plant genera and species to be covered by protection within a period of 10 years. In UPOV 1978, the period of protection ranged from 15 to 18 years. Under UPOV 1991, it ranged from 20 to 25 years. The breeders' rights defined under UPOV 1991 are virtually as tight and comprehensive as the patent rights : the exception for farmers and researchers are very narrowly defined, the paramountcy of the breeders' right being fully protected.

TRIPs logic

In this background, if the government of India continues to assert that, notwithstanding its acceptance of TRIPs agreement, it will not be influenced by the lobbies of the plant breeders and the inexorable logic of the development of higher and higher standards of protection, and will ensure that all the existing rights and practices of farmers remain intact, we should wish it all luck. But we should not forget that when the question was raised in Parliament not long ago on the future of Indian Patents Acts 1970, there was always an assurance that no revision was contemplated. With the TRIPs agreement coming into force in years' time, a total repeal will be the fate of that Act. This experience is too recent to let one feel reassured to lower one's guard on the question of seeds.

SUSTAINABLE USE OF GENETIC RESOURCES

Sustainable use of genetic resources is crucial for the well being of our people. Since there is no streamlined mechanism on this aspect, the resources are being exploited in a very unsustainable manner. In order to develop an effective strategy it is necessary to deal with the following points :

(a) Creation of a national facility for compilation of the scattered information available on our biological resources.

(b) Establishment of a chain of zoological parks, botanical gardens, bio-resource centres, seed gene banks representing biodiversity rich resources and their linkages with three bureaus of the ICAR.

(c) Creation of a central mechanism to deal with administrative policy and regulatory measures for sustainable use of both domesticated and wild variety.

(d) Effective community participation.

So far attention has been paid for the conservation of biodiversity through national parks, sanctuaries, biosphere reserves and other protected areas. However, conservation efforts towards plant species have not been given adequate attention particularly of those which are of potential economic and scientific value. The other important aspect to incorporate is rehabilitative strategy for rate, threatened and endangered plant and animal species. There is further need to develop facilities for long and short-term conservation through :

(1) Establishment of genetic enhancement centres for producing good quality of seeds.

(2) Seed gene banks.

(3) Tissue culture gene banks.

(4) Cryo-preservation.

(5) Pollen storage.

(6) Captive breeding in zoological gardens.

CONSERVATION OF BIODIVERSITY

Man has always been fascinated by the diversity of life. Hunter - gatherers celebrated it through paintings in their caves. Gautam Buddha was born in a secred forest of sal trees and attained enlightenment meditating under a peepal tree (Gadgil, 1992).

People have always evolved many traditions and systems for the conservation of ecosystem. It is is worth mentioning that world's first conservation measures were taken during the time of Emperor Ashok, who set up hospitals and reservoirs for wild animals and birds and extended his protection to many species. Bishnois of Rajasthan emphasises non-violence and respect for all living animals and plants. Chipko movement is one of the most successful examples of people oriented environmental protection. Bishnoi farmers strictly protect every Khejadi tree even in the middle of farmland. Their villages are lush green and alive with blackbuck, chinkara, nilgai and peafowl. Similarly, in Mizoram, tribals have traditionally protected small sacred groves adjacent to village woodlots. They aptly call the groves 'safety forests' and the woodlots 'supply forests'.

The concept of biodiversity conservation is not new to India. The rich cultural heritage shows the presence of various plants in a village and their use by the community. The villagers of Southern Aravallis have been found using more than 100 species of plants. Plants such as *Ficus benghalensis, F. religiosa, Eagle marmelos, Mangifera indica, Helicteres isora, Oscimum sanctum, Asperagus recimosus, Abrus precatorius* etc., are commonly used in performing various retuals and are planted by them in their country yard or farm-yard.

Secred groves are another unique tradition which have been responsible for preserving biodiversity pockets in various parts of the country. Usually a part of temple land, secred groves are areas which are important as they represent the original flora of the locality, preserved in its natural form without outside disturbance. These patches of vegetation and sometimes of the fauna within, for example, snakes and monkeys are considered secred and worshipped by people. Such groves occur in many parts of our country.

The biodiversity conservation programme aims at the conservation of species diversity, genetic diversity and ecosystem diversity found in different regions. The first step in the direction of biodiversity conservation should aim at the conservation of the species diversity. No two species utilize their environment in the similar manner. The existence of different species of organisms therefore means the availability of different alternatives to mankind for utilising their environment in the best possible manner.

The members of the same species growing or living in different climatic and geographical regions sometimes evolve certain genetical properties that distinguish them as a 'variety' of the same species. For example, *Dalbergia sissoo* of Gonda district grow in better shape and size as compared to those in other parts of Uttar Pradeṣh. It is therefore a genetically superior variety. The biodiversity conservation programme should therefore aim not at the conservation of species diversity but also at the genetic diversity existing within the range of the better parent material for higher yield per unit area and also higher resistance against diseases and environmental pressures like drought, water logging and excessive salinity.

The evolution of different genetical properties due to climatic, geographic and environmental influences is an on-going process. The continuance of this evolutionary trend is however, possible only if the undisturbed samples of different ecosystems containing groups of associated communities of plants and animals are preserved in different regions. The conservation of biodiversity and natural ecosystems in different biogeographical regions will go a long way in the preservation of genetic resources and endangered species. At the same time, this would preserve some of our vulnerable ecosystems. That is why, protected areas like National Parks, Wild Life Sanctuaries and Biosphere Reserves are established by different countries over large tracts.

CHALLENGES AND PRIORITIES IN CONSERVING BIODIVERSITY

The current contraction of biodiversity is cause for alarm. And while complete disappearance is most serious, other losses merit attention. Unique races and populations can also be extinguished; such diminishment of genetic variety within a species must be counted a biological loss, for the pool of further evolution is gradually drained (Wolf, 1985).

Biodiversity is related to two main factors. *First*, diversity of species is closely related to the amount of undisturbed habitat available. As a forest is reduced to islands or refuges, many animal species disappear faster than new species arise. *Second*, as populations are localised into smaller fragmented areas, they become vulnerable both to the effects of inbreeding, which diminishes inherent genetic variation, and to chance fluctuations or disease epidemics, that can quickly wipe out small populations.

In addition to human pressures, natural catastrophes regularly disrupt forest lands and can considerably reduce biodiversity. The natural event of recent years that has perhaps had the greatest effect on living diversity was the massive forest fire that blazed for three months in early 1983 in Indonesian Kalimantan on the island of Borneo. Over 3.5 million hectares of forest burned; 800,000 of primary forest, 1.4 million hectares of commercially logged woodland, 750,000 hectares of secondary forest that sustained shifting cultivators, and 550,000 hectares of peat swamps. This charred area, nearly the size of Taiwan, is a single loss equal to nearly two years of human-caused deforestation throughout Southeast Asia (Anon, 1984).

The constellation of factors implicated in Kalimantan's conflagration hold lessons for other tropical regions. In particular, the Brazilian Amazon also hosts large-scale resettlement schemes and timber concessions; a disruption of the area's hydrological cycle that fostered drought conditions could make a similar catastrophic fire possible (Salati and Vose, 1984).

Foundations of Agriculture

The genetic diversity within the handful of crop and livestock species that feed humanity - so called germplasm resources - hold much of the potential for improving agricultural performance. By the early eighties, 10 of the world's 13 international agricultural research centres had begun to focus on germplasm conservation as a vital element of their strategy to boost food supplies. In countries and regions where there are few commercial or national plant breeding efforts, new crop varieties are developed and distributed from the centres. Their efforts to collect and catalogue traditional crop and the wild relatives of such crops in developing countries may be even more important over the long run.

Locally adapted varieties, called landraces, offer an unexpected ocean of potentially useful genes for crop breeders, but unless they are collected when high-performing varieties replace them, they quickly disappear. Botanist Garrison Wilkeys (1983) points out that "the technological bind of improved varieties is that they eliminate the resource on which they are based".

For a decade, collecting and conserving crop germplasm has been coordinated worldwide by one of the 13 international centres. The International Board for Plant Genetic Resources (IBPGR) has overseen field collection of crop varieties and the establishment of gene banks that store at low temperatures the seeds and cuttings of most major food and commodity crops. Though the first and largest gene bank were in the United States, Western Europe, Japan and the Soviet Union, germplasm is now stored in many Third World countries as well (Table 1.2).

Table 1.2. Number of samples of Major Crops held in gene Banks in Industrial and Developing Countries (Plucknett et al., 1983).

Crop	*Industrial Countries*[1]	*Developing Countries*
	Samples (countries)	
Wheat	246,700 (7)	87,000 (3)
Rice	50,800 (3)	148,500 (11)
Maize	40,900 (4)	36,450 (4)
Sorghum	42,900 (3)	37,000 (4)
Barley	127,500 (7)	47,500 (4)
Millet	4,300 (3)	34,500 (2)
Potato	20,600 (4)	21,400 (2)
Soybean	14,350 (3)	15,900 (2)

[1] Includes short-, medium-, and long-term storage facilities.
Source : Adapted from Plucknett et al. (1983).

One specific benefit these gene banks offer the Third World is a measure of insurance against age-old natural and social catastrophes. Stresses like the severe droughts in Africa and regional wars in Central America and Southeast Asia can force people to eat the seed that should be saved for planting. Traditional rice varieties lost in Kampuchea during the seventies, for example, were restored from holdings of the International Rice Research Institute in the Philippines.

Although the establishment of germplasm collections is a positive development, it signals that "the centres of genetic variability are moving from natural systems and primitive agriculture to gene banks and breeders working collections with the liabilities that a concentration of resource (power) implies". One result has been the emergence of political disputes over the control of germplasm.

Controversy erupted around the proposals endorsement of the principle of "free exchange of germplasm". At issue was whether elite breeders stocks, the product of long and costly commercial breeding efforts, should be exchanged among countries on the same basis as wild species and traditional cultivated varieties that had never undergone deliberate scientific selection for their traits. The United States, Japan and countries of Western Europe felt the FAO proposal undermined economic interests and contravened laws that made some breeders stocks proprietary material in their countries. Its advocates including Colombia, Cuba, Libya and Mexico countered that the breeders in developing countries currently had free access to the genetic resources of developing countries that they then developed into commercial varieties to be sold back to the Third World at considerable profit.

There are actually few restrictions on the exchange of germplasm, particularly for breeding purposes, and the growing ranks of gene bank in the Third World should allay fears that the West hopes to "corner the market" on crop germplasm. Many countries urgently need to develop plant breeding programmes to work with stored material and to adapt to national needs from the varieties supplied by the international agricultural research centres. Although private seed companies are seeking market in some developing countries, it is still ministries of agriculture, not private companies that distribute improved seed to farmers in most of the Third World.

The debate on gene banks and who control them tends to obscure the fact that existing collections of major crop are largely complete. Evaluation and use of collected germplasm and an effort to preserve landraces and wild crop relatives where they still exist are the critical needs of germplasm conservation.

CONSERVATION PRIORITIES

The goal of a conservation strategy must be to ensure that evolution continues. Allowing for the play of natural forces by which both wild and domestic species evolve will maintain gene pools and retain genetic traits that may prove valuable in the future. Aside from the biological visdom of protecting species and genes for human needs we needs we cannot anticipate. The world population, now doubling every 40 years, demands that plant breeders select traits that increase yields and harvests. Yet when human populations are stable or growing only slowly, diversity within a crop - a patchwork of varieties - may better suit human needs. The availability of that diversity will depend on conservation choices made now.

Present knowledge about species losses and ecosystem functions, though incomplete, is sufficient to target conservation efforts and to anticipate likely changes. The emerging science of conservation biology, is rapidly enriching our knowledge of loss of biodiversity. A guiding discipline is "conservation genetics", which studies the potential of a species to survive and evolve in parks and managed areas. Scientists can estimate the size of animal populations that will preserve a desired amount of genetic diversity and can forsee biological losses. For example, one study of the population genetics and ecological needs of large animals suggests that even the largest protected areas are unlikely, without intensive management, to sustain viable populations of predators such as the wolf and mountain lion as well as large mammals including elephants, virtually guaranteeing their gradual decline and extinction in the wild within the next century (Frankel, 1984).

Inventories of rare and endangered species can illuminate pressures and suggest preservation priorities. On a global scale, the most important lists have been compiled by the International Union for the Conservation of Nature and Natural Resources. *Red Data Books*, have been published for birds, mammals, amphibians and reptiles, plants and invertebrates. These volumes provide samplings of species known to be endangered in different regions rather than exhaustive lists, guide national and private preservation efforts, IUCN had listed 145 mammals, 437 birds, and 64 amphibians and reptiles from selected groups known to be endangered or threatened in various regions. IUCN also provides data on 250 of a suspected 20,000 - 25,000 threatened plants and over 400 threatened invertebrates to illustrate the pressures on those groups.

A number of countries have taken steps towards listing endangered species, many in response to the Convension on International Trade in Endangered Species of Flora and Fauna. Although this is an important development, few national lists are sufficiently detailed, and fewer countries provide legislative protections as strict as those in Canada, the Soviet Union, or the United States.

Future efforts to protect species and to prevent rare ones from slipping towards extinction will depend on a deeper understanding of the biology of rarity and extinction and a sense of how human interactions with the biosphere affect them.

The success of future efforts to conserve biodiversity rests to a large extent on whether they can be reconciled with development policy. The world Bank took an important step in this direction in 1984 by adopting environmental guidelines for its lending program, the bank has committed itself to refuse to finance economic development projects that will cause irreversible environmental deterioration, including species extinctions (World Bank, 1984). National development efforts should reappraise the value of their remaining wild areas, which are too often considered blank spaces on the map. Hundreds of thousands of hectares of forests have been cleared by farm families participating in massive transmigration and land settlement programmes in Indonesia and Brazil, despite indications that much of this land cannot be continuously cultivated once the forest is cleared.

Biological resources will only be considered when their prises reflect their intrinsic value. The growing demand for genetic diversity in agriculture and the emerging applications of biotechnologies suggest that elusive values will come into sharper economic focus in the years ahead. Evolution has progressed unmanaged for 3.5 billion years, but its future path is certain to be shaped by human forces. In the words of Sir Otto Frankel (1984) "we have acquired evolutionary responsibility". It would be a tragedy if our failure to exercise the responsibility is left for the next Darwin with nothing to write about.

We have numerous and wide ranging policies, programmes and projects, which directly or indirectly serve to protect and conserve the country's biological diversity.

LEGAL FRAMEWORK

Our constitution enshrines the concept of environmental protection. It has laid down important trial in the Section on Directive Principles of State Policy by assigning the duties for the States and all citizens through Article 48A and 51A (g) which state that the "State will endeavour to protect and improve the natural environment including forests, lakes and rivers and wildlife, and to have compassion for the living creatures".

Over time, a legal and policy framework has been developed which relates specifically to biological diversity. On the legal side are the Forest

Act 1927, the Wildlife (Protection Act, 1972, the Forest (Conservation) Act, 1980 and the Environment (Protection) Act, 1986. These are supported by a number of State laws and statues concerning forests and other natural resources.

The National Conservation Strategy, 1992 outlines the policy action required to give greater attention to biodiversity conservation. The National Forest Policy, as amended in 1988, stresses the sustainable use of forests, and on the need for greater attention to ecologically fragile (but biologically rich) areas such as the mountain and island ecosystems. The National Wildlife Action Plan, 1973 lays down the priorities in the areas of Wildlife conservation.

SURVEYS

Surveys of the floral and faunal resources in the country are carried out by the Botanical Survey of India (established in 1890) and the Zoological Survey of India (established in 1916). The Forest Survey of India (established in 1981) uses satellite imagery, aerial photography and ground truth verification to assess the forest and tree cover with a view to develop an accurate data base for planning and monitoring purposes.

IN-SITU CONSERVATION

India has a long history of in-situ conservation of fauna through protected areas. With the setting up of the Indian Board for Wildlife in 1952 and the enactment of the Wildlife (Protection) Act, 1972, the protected areas network has been strengthened. Today, we have 75 National Parks and 421 Sanctuaries covering about 4 percent of the total geographical area of the country.

Project Tiger, launched in 1973, succeeded in increasing the tiger population to more than 4000 in 1989, spread over 19 Tiger Reserves in a total area of 28,000 sq.km. The Wildlife Institute of India has recommended that the number of protected areas should be increased to 148 National Parks and 503 Sanctuaries, covering 4.6 percent of the total area.

The critical problem is not merely the conservation of a particular species or habitat; it is the continuation of the process of evolution itself. In order to ensure the unhindered evolution of micro-organision, plants and animals in their totality, and a part of the natural ecosystems. The

Government of India has already designated 7 biosphere reserves, based on the comprehensive concept of conservation, evolved in 1971 by the UNESCO's Man and Biosphere Reserve Programme. It is proposed to consolidate and expand this programme in the future.

In recent years, programmes have also been launched for ecological research and for protection of fragile ecosystems, including mangroves, wetlands and coral reefs. A national committee on Wetlands, Mangroves and Coral reefs, established by the Ministry, has identified 15 Mangrove areas, 16 Wetlands and 4 Coral reef areas (Gulf of Mannar, Kutch, Andman and Nicobar and Lakshadweep islands) for integrated management and their conservation.

EX-SITU CONSERVATION

To complement the efforts made for in-situ conservation attention has been paid to ex-situ conservation also. There are 33 Botanical Gardens and 33 University Level Botanical Gardens. There are also 205 areas for ex-situ Wildlife Preservation, including 107 Zoos, 49 deer parks, 13 Safari parks, 6 Snake parks, 24 Nature education/breeding centres and 6 aquaria. Some of the major Zoos have made significant achievements in captive breeding of endangered species. For better management of Zoos the Government of India has recently setup the Central Zoo Authority.

Collection and preservation of genetic resources is done through the National Bureau of Plant Genetic Resources, Delhi for the wild relatives of crop plants, the National Bureau of Animal Genetic Resources, Karnal, for the domesticated animals, and the National Bureau of Fish Genetic Resources, Allahabad, for the economically valuable fish species. These bureaus are assigned the task of collecting germplasm from within and outside the country, and also to supply them on request to Indian and foreign agencies for research purposes.

The Department of Biotechnology was established in 1986 to guide, supervise and develop biotechnology programmes in the country and to develop the infrastructure support for this purpose.

RESEARCH AND TRAINING

The Ministry of Environment and Forests is providing financial assistance for several research projects, on identified priority areas of research on environmental conservation and specified ecosystems such as Wetlands, Mangroves and Biosphere Reserves in many Universities and Research Institutions (Singh, 1993).

In the last few years the world has started realising the significance of biodiversity. Most of our food requirement is being met by various plants and living organisms. To meet the demand of future generation, we have to rely largely on better strains of cereals and other plants from the wild. Today out of 80,000 edible plant species, we are using only 30 as major sources of food. It is common knowledge that we require biodiversity contained in plants and animals for food and nutrition, clothing and fabrics, building and shelter, fuel and energy, medicine and biological control and ornamentation and recreation (Pandey, 1991). Resource requirement for medicine and industry in very high which needs to be examined carefully for maintaining the regular supply and balance with the nature. Biological resources are likely to be the basis of all future welfare and security of nations; the sustenance of agriculture, new drugs etc. Biodiversity in a region will be the main strength, power and perhaps the main tool of political bargaining among nations.

REFERENCES

1. Anon. 1984. Wound in the world. *Asiaweek,* July 13.
2. Ehrlich, P.R. 1991. Biodiversity and humanity : Science and Public Policy. *Environmental Awareness,* 14:1:27-35.
3. Fosberg, F.R. 1986. Biodiversity. *Ibid.* 9:4:125-129.
4. Frankel, O.H. 1984. Genetic diversity, ecosystem conservation and evolutionary responsibility. In "*Ecology in practice*". Part I. *Ecosystem management.* Ed. F. Di. Castri et al. Tycooly International Publishing Ltd., Dublin.
5. Gadagkar, R. 1992. World's biodiversity needs to be preserved. *Down to Earth,* 1:11:43-44.
6. Gadgil, M. 1991. Conserving biodiversity: saving subcontinents wealth. *Survey of the environment* 1991. Ed. N.Ravi. The Hindu, Madras. pp. 140-141.
7. Gadgil, M. 1992. Biodiversity: Time for bold steps. *Ibid.* pp. 21-23
8. Kothari, A. 1992. The biodiversity convension:. an Indian viewpoint. *Sanctuary,* 12:3:34-43.

9. Kothari, A. 1993. Biodiversity convension: for those vanishing species. *Survey of the environment 1993*. Ed. N.Ravi. The Hindu, Madras. pp. 44-47

10. Pandey, D.M. 1991. Participatory management and sustainable use of biodiversity. *Yojana*, 35:11-14.

11. Plucknett, D.L. et al. 1983. Crop germplasm conservation and developing countries. *Science*. April 8.

12. Salati, E., and Vose, P.B. 1984. Amazon basin : A system in equilibrium. *Science*, July 13.

13. Senanayake, R. 1993. Is there link between biodiversity and sustainability. *Ecoforum*, 17:3:20-21.

14. Shepard, J.F. et al. 1983. Genetic transfer in plants through interspecific protoplast fusion. *Science*, February 11.

15. Shiva, V. 1992. Biodiversity, biotechnology and profits. *Gandhi Vigyan*, 2:2:47-52.

16. Singh, S. 1993. Conservation of biodiversity in India. Environmental *Awareness*, 16:3:99-103.

17. Wilkeys, G. 1983. *Current status of crop plant germplasm*. CRC Critical Reviews in plant science, 1:2.

18. Wolf, E.C. 1985. Challenges and priorities in conserving biological diversity, *Environmental Awareness*, 8:67-79.

19. World Bank. 1984. *Environmental guidelines*. Office of Environmental Affairs. Washington DC.

RECLAMATION AND REVEGETATION OF BAUXITE AND COAL MINED AREAS IN MADHYA PRADESH

Ram Prasad

INTRODUCTION

Madhya Pradesh is richly endowed with mineral wealth which includes bauxite, coal, manganese, copper, dolomite limestone and iron-ore. Besides clays, various grades of sands and ochre also occur.

The position of the State is more revealing when specific minerals are considered. For example, the State is the only producer of diamond, the largest producer of diaspora (78 per cent), pyrophyllite (48 per cent), dolomite (42 per cent), bauxite (28 per cent) and limestone (23 per cent). The State is the second largest producer of ochre (33 per cent), other sands (29 per cent), iron-ore (24 per cent), coal (13 per cent) and phosporite (14 per cent) (Anon. 1983).

There has been a spectacular growth in the value of mineral production in the state and it is largely contributed by coal. The geographical spread of mineral production is however not uniform, some districts like Bastar, Bilaspur, Jabalpur, Balaghat, Raipur, Durg, Sidhi, Shahdol, Satna, Panna and Jhabua have mineral reserves in a concentrated form while others have restricted to a limited number of pockets. Quite often these mineral reserves are hidden beneath luxuriant forests and the excavation of the mineral reserves brings in its wake large scale destruction of forests and disruption of ecological balance. However, the afforestation trials carried out by the State Forest Research Institute, Jabalpur in various mined out areas have shown that restoration of ecological balance to areas disturbed by mining is possible provided right type of technological inputs are applied.

The State Forest Research Institute, Jabalpur for the first time in 1979, initiated trial plantations of some species for rehabilitation of bauxite mined areas of Balco near Amarkantak. More experiments were taken up in the following years till 1986. In 1982, reforestation trials were started in coal mined areas of SECL at Dhanpuri in Shahdol district. In 1984, some more areas were also reclaimed in coal mined areas at Korba and also in dolomite mined areas of Bhilai Steel Plant at Hirri, both in Bilaspur district. Iron-ore mined out areas of Bhilai Steel Plant at Dalli-Rajahra and limestone areas at Nandini were also taken put in 1984 (Prasad, 1998).

This chapter describes the plantation trials and technological inputs of two important mined areas of bauxite and coal in Shahdol district.

MINERAL WEALTH IN SHAHDOL DISTRICT

Shahdol district is located in the eastern part of the State and is quite rich in mineral resources. Bauxite and coal are the two major minerals being mined, the others being gypsum, sand, ochre, limestone, fireclay and soils. Coal mine belt is stretched in an east-west direction from Bijuri to Umaria. Almost the entire belt has heavy over-burden. Bauxite mines are located at the southern tip near Amarkantak. In selected areas, bauxite has less than 2.5 per cent silica which makes it suitable grade for the aluminium industry. Plenty of sand is available in nala beds and river banks. Ochre occurs in small areas one km. away from Amarkantak; both yellow and red orchres are available.

Limestore deposits are found in Lavehr, Udhan, Bijawar, Larmeta and Kabir. Fireclay is located in Chandia.

A total area of 16,466,262 ha. has been given out on lease for mining purpose in the district of which a huge part. (15,402,120 ha.) is for coal mining. Most of the coal is in Dhanpuri area where the mining lease is for 5,867,847 ha. Thus, the major brunt of land degradation and water and air pollution will be felt in this region (Annon, 1983).

DHANPURI OPENCAST MINES

Opencast coal mining started in Dhanpuri area a little more than a decade back whereas underground coal mining had started in the area some forty years ago. Still there are underground coal mines in area where the coal seam is at greater depths. Dhanpuri area has three opencast mines viz. Amalai, Sharda and Dhanpuri.

The Dhanpuri project has mineable resources to the tune of more than 50 million tones and the targetted output of 1.25 million tonnes. The project has been planned to provide coal to various power stations viz. Amarkantak, Sanjay Gandhi Thermal Power Station and some other consumers of Central and Western India (Annon, 1985).

In the pre-mining stage, the area had profuse forests and some agricultural cultivation near village areas. The area had been rich in wildlife also. As much as 803.43 ha. land belonged to forest department, which is 72 per cent of the total land acquired for mining. Though the actual area leased is 1222 ha. yet the impact is spread over a larger area.

The Dhanpuri project is located in the south-east and south of Amlai colliery in Sohagpur area of SECL in Shahdol district. The PWD road joining Rewa and Amarkantak passes through the project area.

RECLAMATION OF COAL MINED AREAS

The State Forest Research Institute started experimenting through field trials in Dhanpuri since 1982. A number of indigenous and exotic species were tried in different years to identify suitable species and to evolve suitable technology of reclamation (Table 2.1).

Initially block plantations of several species were undertaken. Often monocultures are the basis of successful planting. Under this condition the requirement is both to create as well as maintain the system at an early successional stage. Many fast growing species are pioneers in plant succession (Ewel, 1980).

In the first year (1982), only eight species viz., *Acacia auriculiformis, A. catechu, A. nilotica, Dalbergia sissoo, Emblica officinalis, Pongamia pinnata, Eucalyptus camaldulensis, Eucalyptus* hybrid were tried. Since the top soil of over-burden heaps was totally devoid of humus or organic matter, the pits dug for the planting were replaced with soil brought from the adjoining forests and cow-dung manure. Most of the species were found showing good growth and therefore in subsequent years also the same species were tried. As the technique became perfect, more species viz., *Acacia mellifera, Casuarina equisetifolia, Albizia falcataria, A.*

lebbek, A. procera, Grevillea pteridifolia, Gmelina arborea and *Dendrocalamus strictus* were also tried. In 1984, *Shorea robusta* which is the principal tree species in the natural forests of the area, alongwith its few associates like *Madhuca latifolia* and *Cleistanthus collinus* was also added. Since the existing knowledge of suitable species had been very inadequate the approach adopted by SFRI was to plant a wide range of indigenous and exotic (fast growing) species so as to facilitate selection of suitable species on the basis of growth performance and other values. This process however takes quite a lot of time as the full value is known only after trees attain maturity.

Table 2.1 : List of Indigenous and Exotic Species tried on coal mine overburdens

Indigenous	*Exotic*
NFT :	NFT :
Acacia catechu	*Acacia auriculiformis*
A. nilotica	*A. albida*
Dalbergia sissoo	*A. mellifera*
Pongamia pinnata	*A. tortilis*
Albizia lebbek	*A. victoria*
A. procera	*Albizia falcataria*
	Gliricidia sepium
NON NFT :	NON NFT :
Bombax ceiba	*Eucalyptus camaldulensis*
Cleistanthus collinus	*E. citriodora*
Emblica officinalis	*E.* hybrid
Holoptelia integrifolia	*Grevillea pteridifolia*
Gmelina arborea	*Casuarina equisetifolia*
Dendrocalamus strictus	
Shorea robusta	
Terminalia belerica	

Over a period of one decade, the growth performance of these species reveals that both *Eucalyptus* species *(E. camaldulensis* and

Eucalyptus hybrid) were the most successful, both in terms of height as well as girth as shown in Tables 2.2 & 2.3 (Prasad and Shrivastava, 1991). Indigenous species such as *Acacia catechu, A. nilotica* and *Dalbergia sissoo* have also performed satisfactorily.

Table 2.2 : Trend of height increase (cm) in various species tried at coal mine overburdens

Experiment started since 1982 *Observations in December*

Species	*1984*	*1986*	*1988*	*1990*	*1992*
Acacia auriculiformis	224	336	581	600	640
Acacia catechu	257	348	579	734	740
Acacia nilotica	281	394	512	607	700
Dalbergia sissoo	174	292	378	440	480
Emblica officinalis	127	168	284	357	394
Eucalyptus camaldulensis	504	680	889	1057	1140
Eucalyptus hybrid	548	720	1033	1195	1580
Pongamia pinnata	229	299	365	441	550

Table 2.3 : Trend of girth increase (cm) in various species tried at Coal mine overburdens

Experiment started since 1982 *Observations in December*

Species	*1984*	*1986*	*1988*	*1990*	*1992*
Acacia auriculiformis	9.5	13.7	21.7	30.2	32.0
Acacia catechu	7.2	13.9	18.0	23.9	24.8
Acacia nilotica	6.4	10.5	13.4	17.2	19.0
Dalbergia sissoo	6.2	11.8	17.0	24.0	25.0
Emblica officinalis	4.6	9.5	10.7	12.8	13.5
Eucalyptus camaldulensis	13.6	21.6	26.0	33.9	35.0
Eucalyptus hybrid	12.8	22.8	25.3	38.2	40.0
Pongamia pinnata	4.8	10.2	11.3	15.8	16.0

Table 2.4 : Natural regeneration of different species tried at coal mined areas

As in - Dec. 1990

	Year of Plantation			
Species	*1982*	*1983*	*1984*	*1985*
		No. of Plants per ha.		
A. auriculiformis	3,30,000	2,60,000	3,60,000	2,60,000
P. pinnata	30,000	-	-	-
E. hybrid	50,000	-	-	-
E. camaldulensis	-	18,000	60,000	-
G. pteridifolia	-	-	-	-
A. catechu	2,000	-	-	10,000

In order to study the natural regeneration in plantations of different years, survey of ground flora was done. The data revealed that *Acacia auriculiformis* had the best natural regeneration (Table 2.4). Under monoculture, the maximum recruits were found under *A. auriculiformis, E. camaldulensis* and *E.* hybrid.

BIOMASS PRODUCTIVITY

Mean trees representing the average height and diameter at breast height of the selected successful species were selected and felled. Bark, twigs, roots, trunk and branches were weighed separately to take the green weight. The biomass samples were then oven-dried and weighed again. On the basis of actual survival per cent, the per ha. productivity of six species viz., *Eucalyptus camaldulensis, Eucalyptus* hybrid, *Pongamia pinnata, Acacia auriculiformis, Acacia nilotica* and *Acacia catechu* was calculated. The data revealed that in terms of above ground as well as below ground biomass of 10 years old plantation, the two eucalyptus viz., *Eucalyptus* hybrid and *E. camaldulensis* the most successful species, produced 150.922 t/ha. and 144.20 t/ha. above ground and 22.53 t/ha. and 32.23 t/ha. below ground biomass respectively. This was followed by *Acacia auriculiformis* (88.680 t/ha. and 14.94 t/ha.) and *A. catechu* (52.91 t/ha. and 23.44 t/ha.). Among these six species, *A. nilotica* proved to be the poorest as it gave only 27.34 t/ha. above ground and 2.73 t/ha. below ground biomass (Table 2.5).

Table 2.5 : Comparative Biomass Efficiency of Selected Species Planted in 1982 on coal mine overburden areas

Particulars	*Eucalyptus* hybrid	*E. camaldulensis*	*P. pinnata*	*A. auriculiformis*	*A. nilotica*	*A. catechu*
No. of tree/ha	1333	2066	2250	1200	1100	2150
Mean tree ht (m)	15.80	11.40	5.50	6.40	7.00	7.40
Mean tree						
Over bark	40	35	16	32	19	24.8
d.b.h. (cm)						
Under bark	36	30	14	28	16	22.0
Mean tree biomass (kg.)						
(a) Wood	98.70	59.75	15.70	62.00	22.70	22.60
(b) Bark	8.52	4.50	4.50	5.50	1.50	1.74
(c) Leaves	6.00	5.55	0.35	6.40	0.65	0.27
Total above grounds (kg.)	113.22	69.80	20.55	73.90	24.85	24.61
Below grounds (kg.)	16.90	15.60	5.10	12.45	2.50	10.90
Total plant biomass (t/ha.)						
(a) Above grounds	150.90 (15.09)	144.20 (14.42)	46.24 (4.62)	88.68 (8.87)	27.34 (2.73)	52.91 (5.29)
(b) Below grounds	22.53 (2.25)	32.23 (3.22)	11.48 (1.15)	14.94 (1.49)	2.75 (0.28)	23.44 (2.34)
Leaf litter Production (t/ha.)	12.88	14.81	14.02	27.45	0.99	3.78

(Figures in parenthesis are t/ha./vr MAI)

Thus, *Eu-calyptus* hybrid and *E. camaldulensis* are the most suitable species for coal mined areas, particularly in the initial period. However, in view of comparatively poor canopy cover, slow enrichment of soil and lack of natural regeneration, this species is not an ideal species and should be planted only when biomass production is the foremost consideration. On sites like mine O.B. dumps, merely biomass production can not determine the choice of species but ecological and socio-economic considerations play more dominant role.

CANOPY COVER

For studies on canopy cover, ten promising species were repeated in different plantations from 1982 to 1986. The canopy cover was assessed by measuring the average spread of canopy from tree axis. Mean spread at right angle to the tree was assessed.

Table 2.6 : Canopy Studies of various species tried at Coalmine Overburdens at different plantation years

Crown cover (sq. m)
Observation in February 1992

S. No.	*Species*	*1982*	*1983*	*1984*	*1985*	*1986*
1.	*Acacia catechu*	1.02	0.97	0.97	0.84	0.62
2.	*Pongamia pinnata*	1.67	1.63	1.72	-	-
3.	*Eucalyptus* hybrid	1.22	1.11	1.15	0.99	0.67
4.	*Acacia auriculiformis*	2.12	1.97	1.92	1.85	1.85
5.	*Emblica officinalis*	-	-	-	0.84	-
6.	*Eucalyptus camaldulensis*	0.97	0.77	0.76	0.62	-
7.	*Dalbergia sissoo*	0.67	0.61	0.49	-	-
8.	*Holoptelia integrifolia*	-	1.92	3.45	-	-
9.	*Shorea robusta*	-	-	0.06	-	0.03
10.	*Grevillea pteridifolia*	-	-	-	-	1.09

The data given in Table 2.6 shows that after 10 years (1982 plantation). *Acacia auriculiformis* provided the thickest canopy cover (2.12), followed by *P. pinnata (1.67)*, *Eucalyptus* hybrid (1.22) and *Acacia catechu* (1.02). Among the poorest were *E. camaldulensis* (0.97) and

Dalbergia sissoo (0.67). Indigenous species *Shorea robusta* has been found with the thinnest canopy cover because it is a slow growing species. On the other hand *Grevillea pteridifolia* even after 3-4 years of age develops thicker canopy (Anon, 1993).

Bamboo *(Dendrocalamus strictus)* plantation was also undertaken in the year 1982, 1983, 1984, 1985 and 1986 on the O.B. dumps. Growth performance of bamboo was found to be better on slopes of overburden dumps. Recruitment of new culms enhanced the canopy cover and thus helped in reducing soil erosion. Mean circumference of bamboo culms was found to be 2.70 m in ten years. 2.38 m in nine years. 1.84 m in six years and 1.26 m in four year old plantations. There were as many as 31 culms in a clump in the ten year old plantation. The maximum culm height was found to be 3.39 m and basal girth 8.26 cm after six years of growth.

TECHNOLOGY EVOLVED

The main objective of revegetation of mined out area was to stabilise the dumps through vegetative cover and help build a self-sustaining ecosystem. The approach adopted by the State Forest Research Institute was to treat the overburdens in their existing shape. Initially those species were tried which had shown tolerance to derelict sites. Since the available soil was highly nutrient deficient it was supplemented with some forest soil and manure. so that in the initial period the saplings received adequate nutrients for rapid establishment and growth.

Since selection of suitable species is extremely important to the development of a proper self-sustaining ecosystem, several species, indigenous and exotics, were tried with varying degree of success. A number of parameters viz.. growth survival. litter productivity. litter decomposition, canopy cover, nitrogen fixation, soil enrichment, etc. were taken into consideration to select site-friendly species.

On the basis of growth performance *Eucalyptus* sp., *A. auriculiformis. A. catechu.*, *G. pteridifolia*, *Dendrocalamus strictus*, *D. sissoo. Albizia lebbek*, *A. procera* and *A. mellifera* were found suitable species. keeping other parameters like good regeneration, establishment, nitrogen fixation, canopy, litter productivity, etc. also in consideration, *A. auriculiformis*, *G. pteridifolia*, *P. pinnata*, *D. sissoo. A. nilotica* and *A. mellifera* have been found suitable species.

Though *Eucalyptus* sp. have shown extremely good performance in growth and biomass production, its performance in encouraging growth of associates and ground flora has been dismal. Decomposition of leaves is also rather slow. In practice., we would prefer to recommend only those species which encourage development of other species including ground flora. Eucalyptus is at best a monoculture species.

It is beyond doubt that an integrated approach covering slope stabilisation, soil and moisture conservation, terracing, benching and mulching, combined with suitable species selection, is the best method of reclamation.

In overburdens where the slope exceeded 20° , and where planting was not successful, earthen and biological contour trenches have ensured success in plantation, moisture retention and stabilisation of slopes. In the slopes *Aqave, Eulaliopsis binata. Prosopis iuliflora and A. nilotica* (thickly sown) have helped to stabilise and prevent erosion.

On the basis of trials, it has been found that one year old saplings were most suited for planting on sites. which were levelled and prepared. Forest soil (Sal forests) in pits gave good results probably because of the presence of mycorrhizal inoculum in the soil.

Experiments have shown that 30 cm^3 pits are appropriate for plantation in overburden dumps. In O.B. dumps, soil is loosely stacked and has all the favourable attributes to provide better aeration and space for root growth. Thus pit size is not of important consequence. On escapement., it has been observed that spacings of 2 m x 2 m and 2 m x 1 m (in case of extremely loose soil) have been found to be most appropriate.

BAUXITE MINED AREAS

Reforestation trials in bauxite mined areas of BALCO near Amarkantak were started in 1979 and by 1986., more than two dozen species indigenous as well as exotic, had been tried and 50.60 hectares area had been planted up (Table 7). The climate of Amarkantak is comparatively cooler with mean maximum and mean minimum temperatures as 32°C and 1°C, respectively. Mean annual rainfall varies from 1400 mm to 1750 mm.

Table 2.7 : List of Indigenous and Exotic Species tried at Bauxite Mined Areas

Indigenous	*Exotic*
Albizzia procera	*Acacia auriculiformis*
Cleistanthes collinus	*Albizzia falcataria*
Dalbergia sissoo	*Eucalyptus camaldulensis*
Dendrocalamus strictus	*Eucalyptus citriodora*
Emblica officinalis	*Eucalyptus grandis*
Holoptelia intigrifolia	*Eucalyptus* hybrid
Melia azadirach	*Grevillea pteridifolia*
Pongamia pinnata	*Grevillea robusta*
Shorea robusta	*Pinus caribaea*
Sterculia urenus	*Pinus kesiva*
Toona ciliata	*Pinus oocarpa*
Kydia claycina	*Pinus roxburghii*
	Melaleuca leucodendron
	Moringa species

The original forest vegetation of the area can be classified as moist peninsular high level sal (3C/C2). Sal (Shorea robusta) has been the main constituent of the top canopy (60.80%), tending to be pure on slopes and on plateaus. The general crop density on slopes and in valleys varies from 0.7 to 0.9.

In the mined area the shrubby understorey and ground vegetation have been destroyed on account of excessive biotic pressure.

During mining, the standing trees are removed and the ore is extracted by digging large pits. After extraction of the ore, the mine overburden is heaped in the form of dumps. Although these dumps are roughly levelled with the help of dozers by mining authorities before planting, the top surface, mainly consisting of lateritic boulders, remains undulating. Surface soil is acidic with pH varying from 5.5 to 6.5

Table 2.8 : Comparative growth performance of different species at the age of 10 years tried at the Bauxite Mine overburdens

Species	*Year of planting*	*Survival percent*	*Mean Height (cm)*	*Mean gbh (cm)*
A. Indigenous Species				
1. *Pongamia pinnata*	1979	82	217	6.0
2. *Shorea robusta*	1979	78	224	10.2
3. *Emblica officinalis*	1982	94	107	6.2*
4. *Sterculia urenus*	1982	44	86	15.6*
5. *Albizzia procera*	1983	91	103	7.8*
6. *Cleistanthus collinus*	1983	39	184	19.0*
B. Exotic Species				
1. *Acacia auriculiformis*	1979	74	462	14.0
2. *Eucalyptus camaldulensis*	1979	92	1123	25.3
3. *Eucalyptus hybrid*	1979	94	1012	23.9
4. *Grevillea pteridifolia*	1979	93	670	24.3
5. *Grevillea robusta*	1979	91	565	18.3
6. *Pinus caribaea*	1979	85	713	31.4
7. *Pinus roxburghii*	1979	95	342	26.3
8. *Pinus oocarpa (FRI/625)*	1980	72	769	26.9
9. *Pinus oocarpa (FRI/655)*	1980	77	639	32.5
10. *Eucalyptus grandis*	1981	86	843	22.9
11. *Pinus cassia*	1981	33	655	53.4
12. *Eucalyptus citriodora*	1983	81	1040	28.4

* Shows collar girth instead of girth at breast height (gbh).

PLANTING TECHNIQUE

Soil pits of size 45 cm^3 were dug at a spacing of 2× 2 m. These pits were refilled with surface soil brought from expected, their performance in general, has been much poorer as compared with that of the fast growing exotics. Among the 12 species tried. 3 species viz., *Melia azadirach. Toona ciliata* and *Dendrocalamus strictus* failed. *Albizia procera, Dalbergia sissoo* and *Kydia claycina* have been tried in the later years and their initial performance is promising.

Among the other six indigenous species, for which the comparative growth data of 10 years is available, the performance of Sal *(Shorea robusta)* with mean height of 2.24 m and mean gbh 10.2 cm at the age of 10 years, has been quite promising. The other successful indigenous species are *Cleistanthus collinus* and *Ponoamia pinnata.*

In terms of overall performance, the successful species on bauxite mined sites are *E. citriodora, E. camaldulensis, P. caribaea, P. oocarpa, E.* hybrid *G. pteridifolia* and *G. robusta.* In terms of biomass production *A. auriculiformis* was found to be performing better than *P. roxburghii, Shorea robusta* and *P. pinnata.*

BIOMASS PRODUCTIVITY

Five trees (representing mean trees) of each species were selected from 1979 trial plantation and biomass measurements were taken separately for wood, bark, leaves and roots of these trees. The comparative production performance of selected species in 5 year old plantation is given in Table 2.9.

Total above ground biomass of 110.951 t/ha and below ground biomass of 6.510 t/ha was observed in case of *Eucalyptus camaldulensis.* In terms of mean annual increment, it comes to about 23.47 t/ha/yr biomass production. The two species *G. pteridifolia* and *P. caribaea* produced almost identical biomass (87.551 and 81.263 t/ha, respectively). Another exotic *G. robusta* was found to be next successful species with 73.341 t/ha of above ground biomass.

In terms of litter production, *G. pteridifolia* has been found to be most effective producer with 16.624 t/ha litter. Production of leaf litter from *E. camaldulensis* plantation (10,475 t/ha), is also good. Among the successful species of trial plantations of 1979, a large number of experiments on pit size, spacing, interplanting of Sal, different types of soil refill in the pits, manuring and fertiliser treatments have also been taken up.

Table 2.9 : Comparative production performance of successful species on bauxite mined areas

Planting : 1979 — *Observation : December, 1984*

S. No.	*Attributes*	*Species*				
		Eucalyptus camaldulensis	*Acacia auriculiformis*	*Grevillea robusta*	*Grevillea pteridifolia*	*Pinus caribaea*
1	*2*	*3*	*4*	*5*	*6*	*7*
1.	No. of trees/ha.	2325	2250	2450	2450	2325
2.	Mean tree height (m)	9.83	3.50	3.85	6.30	4.47
3.	Mean tree dia at breast ht.	7.5	2.4	5.2	7.1	6.9
4.	Mean tree volume (cm)					
	a. Over bark	0.024095	0.001919	0.01182	0.01262	0.014583
	b. Under bark	0.18650	0.001291	0.00825	0.01072	0.008197
5.	Total volume (cu m/ha.)					
	a. Over bark	56.021	4.318	28.959	30.919	33.905
	b. Under bark	43.361	2.905	20.213	26.264	19.058

6.	Mean over dry tree biomass (kg.)					
	a. Wood	33.802	2.432	21.055	25.930	17.275
	b. Bark	6.670	0.570	2.350	2.270	2.205
	c. Leaves	7.206	1.000	2.250	2.735	9.472
	d. Total above ground biomass	47.678	4.002	25.655	30.935	28.952
	e. Below ground biomass	2.800	0.800	4.280	4.800	6.000
7.	Total plant biomass (t/ha.)					
	a. Above ground	110.851	9.005	62.855	75.791	67.313
	b. Below ground	6.510	1.800	10.486	11.760	13.950
8.	Total leaf litter accumulation (t/ha.)	10.475	4.150	8.450	16.624	1.997

The experimental plantations raised at bauxite mined areas have clearly exhibited that these areas can be successfully reclaimed by planting suitable species and by adopting the right technology.

REFERENCES

1. Anon., 1983. Environmental status of mining in Madhya Pradesh. An overview. Environmental service group. World Wildlife Fund, New Delhi.
2. Anon., 1985. Environmental Management Plan for Dhanpuri opencast project. Central Mine Planning & Design Institute, Nagpur.
3. Anon., 1993. Studies into ecological and socio-economic impact of opencast coalmining and evolution of technology for restoration of overburden heaps, Project Report, State Forest Research Institute, Jabalpur.
4. Ewel. J. 1980. Tropical succession, manifold routes to maturity. *Biotropica, 12* (Supplement). 2-7.
5. Prasad, Ram and J.L. Shrivastava 1991. Most successful species on coal O.B. dumps. *Vaniki Sandesh,* Vol. XV(2) : 1-4.
6. Prasad, Ram 1988. Technology of Wastelands Development, Associated Publishing Co., Karol Bagh, New Delhi.

ENVIRONMENTAL CHANGES AND ITS IMPACT ON BIODIVERSITY AT AND AROUND PACHMARHI HILLS OF MADHYA PRADESH

M. Oomachan

ABSTRACT

This chapter deals with the impact of environmental changes on Flora and Fauna at and around Pachmarhi hills including its forest area. The observations made for the last three decades showed there is great decline in the biodiversity and a good number of the rare flora and fauna of this region is threatened and are vanishing. Few of them which were reported earlier have been fully eliminated due to intense bio-edaphic interferences and consequent environmental changes.

INTRODUCTION

Pachmarhi and its adjacent forest reserves in Sohagpur tahsil of Hoshangabad district of Madhya Pradesh is situated in 22° 28' North latitude and 78° 26' East longitude on a plateau of the Satpura range, 52 km. from Pipariya Railway Station. Pipariya is on the Itarsi-Allahabad section of Central Railway. Pachmarhi is well connected with Pipariya by road and the M.P. State Roadways buses ply regularly.

Pachmarhi is a lovely hill station girdled with panoramic plateau with great scenic beauty and is on the evergreen Satpura range, and is therefore popularly known as the "Queen of Satpura". Every year many batches of botanical tour parties visit this place from different parts of the country and even from abroad for the nature study and collection of a number of rare botanical specimens and hence it is referred as "Botanists' Paradise".

The name 'Pachmarhi' is believed to be derived from the "Panch Marhi" or the five caves of the Pandava brothers still exist there. These five ancient dwellings excavated in the sandstone rock in a low hill are now protected monuments. Pachmarhi is having the highest mountain peaks in Central India, such as Dupgarh (1350m) on the South-West, Mahadeo (1336m) on the South and Chauragarh (1316m) on the South East directions. The altitude of this region is 1067 metres and average rainfall is about 200 cm. The ranges of temperature during the months of May-June is 39°C to 19°C respectively.

A good number of workers made checklists of the Flora and Fauna of this region. Forsyth in his "Highlands of Central India" has given certain details of the animals and plants characteristic of this region. He has mentioned about the bison (*Bos gaurus*) he found grazing by the huge tree which lies between what is now known as Bison Lodge and Rock End Bungalow. Bison was found in the well known open glade to the north of Handi Khoh. Other animals usually observed in the past were Jackals, Barking deer, Sambhar, Wild dog, Leopard, Tiger, Ruddy mongoose, Indian porcupine, Striped hyaena, Black-naped hare, Otter, Bengal monkey, Sloth bear, Common Striped Squirrel, Large Indian Squirrel, Langur, Indian Wild Boar, Four-horned Antelope, Madras Tree-shrew and Indian Fox. About 120 different species of birds are found in the vicinity of Pachmarhi. Insects and Reptilia are enumerable. Large trees, form the upper storey with economically important timber species. The mid-storey is composed of medium-sized trees, shrubs and climbers. The Lower or under-storey has many valuable medicinal and other useful herbs. The deep shady valleys abound in rare ferns and bryophytes. The Algae, Fungi and Lichens are also found in good numbers.

MATERIALS AND METHODS

The Flora of this region received special attention of many workers. Important among them are Mukherjee, Narayanaswamy and Rao, Kapoor and Yadav, Panigrahi *et al.*, Oommachan, Oommachan and Masih, Oommachan *et al.* Oommachan and his students made continuous observations on the Flora of Pachmarhi since 1964 and published a number of papers. As a result of many trips made during the last over 3 decades Oommachan (1992) made 156 additions to the Angiospermic Flora of Pachmarhi and Bori reserves. Field notes on medicinal plants, rare and endangered species, ethnobotanically important plants and also list of vanishing species prepared. The collections of plant materials are preserved in the herbarium and museum in the Department of Biological Science, Rani Durgavati University, Jabalpur and also at the Herbarium of the State Forest Research Institute, Polipathar, Jabalpur for future reference.

RESULTS AND DISCUSSION

The Flora and Fauna of Pachmarhi has been largely removed from the valley regions due to the felling, clearing and extensive cultivation. So in many areas what we get today is only the remnants of the past climax vegetation or are otherwise the bio-edaphic communities. Studies made by the present worker during 1964 to 1993 showed that there is continuous decline of bio-diversity in Pachmarhi.

More than 1000 vascular plant species are reported, out of which more than 10% is at the verge of extinction and are rapidly vanishing. More than 30 species, as given in Table 3.1, have vanished or are untraceable which were present in the past. The main reason for their elimination is the irrational exploitation of the economically useful and medicinal plants for economic gains without any thought of its conservation. Further, it is observed that botanical and drug plant collection parties also waste or destroy certain rare plants of this area. Due to the various bio-edaphic changes of the area, including drastic climatic changes, a number of new introductions are also observed (Oommachan, 1992). Table 3.1 shows certain Himalayan species still found in Pachmarhi, South and South-West Indian Species, Certain rare Pteridophytes, vanishing/untraced species, and an exhaustive list of rare plants of Pachmarhi. Many species of animals too have disappeared now. Careful management of this very important hill station and its surrounding forest biodiversity warrant our immediate attention for its conservation and protection.

Table 3.1: Certain Rare Plants of Pachmarhi

A. Himalayan Species:

1. *Rubus ellipticus* J.E. Sm: "Hinsalu" Rosaceae.
   Fl. & Fr. : Dec. - Jan. Locality : 'Kua Khad'.
2. *Berberis asiatica* Roxb. ex DC. "Chitra" Berberidaceae.
   Fl. & Fr. : March - June. Locality : Pachmarhi hills.
3. *Thalictrum foliolosum* DC. 'Pilazari' Ranunculaceae.
   Locality : Mahadeo hills.

4. *Centranthera nepalensis* D. Don. Scrophulariaceae.
5. *Viscum nepalense* Spreng. 'Bandala', 'Posibanda' Loranthaceae.

B. South and South-West Indian Species:

6. *Curcuma pseudomontana* Graham Zingiberaceae.
   Locality : Pachmarhi hill slopes. Fl. & Fr. : July - December.
7. *Begonia malabarica* Lamk. Begoniaceae.
8. *Nilgirianthes campanulatus* (Wt.) Bremek Acanthaceae
   Pachmarhi & Bori Reserve forest. Fl. &Fr. : July-December.
9. *Rostellularia latispica* (Clarke) Bremek Acanthaceae.
   Fl. & Fr. Sept. - March. Locality : Forest undergrowth.

C. Rare Pteridophytes:

10. *Psilotum triquetrum*
11. *Isoetes panchanenii*
12. *Selaginella exigua*
13. *Ophioglossum nudicauli*
14. *Polybotrya appendiculata*
15. *Cyathea gigantea*
16. *Cyathea spinulosa*
17. *Alsophila glabra*

D. Vanishing/Untraced species:

1. *Alysicarpus glumaceus* (Vahl) DC
2. *A. hamosus* Edgew.
3. *Atylosia scarabaeoides* (L.) Baker
4. *Crotalaria retusa* L.
5. *C. trifoliastrum* L.
6. *C. triquetra* Dalz.
7. *Desmodium rotundifolium* Baker
8. *D. velutinum* (Willd.) DC.
9. *D. dichotomum* (Willd.) DC.
10. *D. heterocarpon* (L.) DC.
11. *Indigofera astragalina* DC.

12. *Teramnus mollis* Benth.
13. *Vigna trilobata* (L.) Verdc.
14. *V. vexillata* (L.) A. Rich.
15. *V. mungo* (L.) Hepper
16. *V. radiata* (L.) Wilezek
17. *V. dalzelliana* (O.K.) verdc.
18. *V. umbellata* (Thunb.) O.&O.
19. *V. unguiculata* (L.) Walp.
20. *Mucuna* pruriens (L.) DC.
21. *Sesbania bispinosa* (J.) W.F.W
22. *Albizia procera* (Roxb.) Benth.
23. *Dichrostachys cinerea* (L.) Wt.
24. *Mimosa himalayana* Gamble
25. *Baliospermum Montanum* (W.) M.A.
26. *Euphorbia heterophylla L.*
27. *Homonoia riparia Lour.*
28. *Boehmeria platyphylla* D.Don
29. *Girardinia zeylanica* Decne.
30. *Ficus exasperata* vahl.

E. List of rare plants of Pachmarhi which need protection

1. Cochlearia cochlearioides (Roth) S. & M. Brassicaceae
2. Viola betonicifolia J.E. Smith Violaceae
3. Cochlospermum religiosum (L.) Alston. 'Kumbhi' Cochlospermaceae
4. Viccaria pyramidata Medik. 'Musna' Caryophyllaceae
5. Hibiscus panduraeformis Burm.f, Malvaceae
6. Geranium mascatense Boiss. Geraniaceae

7.	Biophytum peteresianum Klotzsch. Oxalidaceae	
8.	Crotalaria nana Burm.f. Fabaceae	
9.	Potentilla supina L. Rosaceae	
10.	Sonerita tenera Royle. Melastomaceae	
11.	Rotala mexicana Cham. & Schecht. Lythraceae	
12.	Corallocarpus epigaeus (Rottl. & Willd) Clarke Cucurbitaceae.	
13.	Bupleurum plantaginifolium Wt.	Apiaceae
14.	Pycnocycla glauca Lindl.	-do-
15.	Sanicula elata Buch. Ham. ex D. Don.	-do-
16.	Trachyspermum stictocarpum var. hebecarpum (Cl.) Wolff	-do-
17.	Hymenodictyon excelsum (Roxb.) Wall. 'Bhaulan' Rubiaceae	
18.	Conyza leucantha (D.Don) L. & R.	Asteraceae
19.	Galinsoga parviflora Cav.	-do-
20.	Gnaphalium luteo-album L.	-do-
21.	Senecio bombayensis Balakr.	-do-
22.	Sonchus brachyotus DC. 'Sahadevi bari'	-do-
23.	Synedrella nodiflora (L.) Gaertn.	-do-
24.	Stylidium tenellum Sw.	-do-
25.	Campanula benthamii Wall. ex Kitamual	Campanulaceae
26.	Anagalis pumila Sw.	Primulaceae
27.	Lysimachia candida Lindl.	-do-
28.	Leptadenia pyrotechnica(Forsk.) Decne.	Asclepiadaceae
29.	Buddleja medagascariensis Lamk.	Loganiaceae
30.	Solanum incanum L.	Solanaceae
31.	S. seaforthianum Andr.	-do-

32. Kickxia incana (Wall.) Pennell	Scrophulariaceae
33. Limnophila aromatica (Lamk.) Merr.	
34. L.connata (Buch.Ham. ex D. Don) Hand	-do-
35. Lindernia hookeri (Clarke) Wettst.	-do-
36. L. hyssopoides (L.) Haines	-do-
37. L. procumbens (Krock) Phileox.	-do-
38. Utricularia exoleta R. Br.	Lentibulariaceae
39. Hygrophila polysperma (Roxb.) T. Anders	Acanthaceae
40. Holmskioldia sanguinea Retz. 'Kapni'	Verbenaceae
41. Lamium amplexicaule L.	Lamiaceae
42. Leucas zeylanica (L.) R. Br.	-do-
43. Plectranthus rugosus Wall.	-do-
44. Glochidion johnstonei Hook.f.	Euphorbiaceae
45. Aerides multiflora Roxb.	Orchidaceae
46. Eulophia explanata Lindl.	-do-
47. E. herbacea Lindl. 'Ban singara'	-do-
48. Malaxis mackinnonii (Duthie) Ames	-do-
49. Nervilia aragoana Gaud. 'Ban batasha'	-do-
50. Rhynchostylis retusa (L.) Bl. Bijdr.	-do-
51. Dioscorea wightii Hook.f.	Dioscoreaceae
52. Cyperus diaphanus Schr. ex Roem. & Schu	Cyperaceae
53. Brachiaria deflexa (Schu.) C.E. Hubb ex Robyns	Poaceae
54. Dimeria connivens Hack.	-do-
55. Isachne gracilis C.E. Hubb.	-do-
56. Ischaemum duthiei Stapf ex Bor.	-do-
57. I. semisagittarum Roxb.	-do-
58. Manisuris clarkei (Hack.) Bor.	-do-
59. M. forficulata C.E.C. Fischer	-do-

60. Oropetium roxburghianum(Steud.) S.M. Phillips	-do-
61. Paspalum distichum L.	-do-
62. P. orbiculare G. Forst.	-do-
63. Rhynchelytrum villosum (Parl.) Chiov.	-do-
64. Schizachyrium exile (Hochst.) Pilger	-do-
65. Tripogon lisboae Stapf.	-do-

ACKNOWLEDGEMENT

The Author is grateful to Professor S.K. Hasija, Head, Deptt. of Biological Studies, Rani Durgavati University, Jabalpur for providing necessary facilities for this investigation. He is also indebted to the Principal, the convener and other Faculty members of K.G. Arts and Science College, Raigarh for inviting this article for publication in the Proceedings of the "Research Seminar on Environmental Changes, its impact on Flora and Fauna".

REFERENCES

1. Kapoor, S.L. and H.L. Yadav. 1962. Further contribution to the Flora of Pachmarhi Region. Indian For. 88:272-276.

2. Mukherjee, A.K. 1984. Flora of Pachmarhi and Bori Reserves. B.S.I. Howrah. 407 pp.

3. Narayanaswamy, V. and R.S. Rao. 1960. A contribution to our knowledge of the vegetation and flora of the Pachmarhi plateau and the adjacent regions. J. Indian Bot. Soc. 39:222-242.

4. Panigrahi, G., C.M. Arora, D.M. Verma and V.N. Singh, 1966. Contribution to the Botany of Madhya Pradesh-1 (Dilleniaceae to Moringaceae. Bull. Bot. Surv. India. 8:117-125.

5. Panigrahi, G.M. and C.M. Arora.1965. Contribution to the Botany of Madhya Pradesh-II (Rosaceae to Rubiaceae) Nat. Acad. Sci.B35,Pt.I:87-98.

6. Panigrahi, G., C.M. Arora and D.M. Verma.1965. Contribution to the Botany of Madhya Pradesh-III (Ebenacea to Convolvulaceae). Nat. Acad. Sci.India. Sec.B.35, Pt.1:99-109.

7. Panigrahi, G. and R. Prasad.1967. ibid-iv-Euphorbiaceae and Urticaceae.ibid.B.37,Pt.4:553-564.

8. Oommachan, M. 1992. Additions to the Flora of Pachmarhi and Bori Reserves (Madhya Pradesh).Indian J. Applied and Pure Biol.7(2):71-75.

9. Oommachan, M. and S.K. Masih. Contribution to the Flora of Pachmarhi - A Reassesment. J. Econ. Tax. Bot. 16 (2):437-445.

AIR QUALITY AND THERMAL ENVIRONMENTAL CHANGES UNDER TROPICAL CONDITIONS OF URBAN HYDERABAD – A CASE STUDY

S.H. Raza

ABSTRACT

Building activity in the urban habitats have recently increased with increase in city dwellers. This has resulted in the formation of slums and industrial urban centres. Vegetational areas have been drastically reduced. This paper deals with the comparative account of the thermal environment and the air quality of the urban-industrial area and the green belt area. There has been 2-3°C rise in temperature in the urban agglomerations forming heat islands in the heart of the city. Planting of green belts and the water bodies has reduced the temperature from 7-11°C between 12 noon to 4 pm in these vegetated zones. Suitable ability index (SAI) of the plants have been derived. Plants having higher SAI grade can be suitably used to create cooling effect and to improve the air quality of the area. Some of the species are *Mangifere indica. Artocarpus integrifolia. Ficus benghalensis. F. infectoria. Memusops elongi. Dalbergia sissoo. Shizigium* sp. having higher SAI grade 7, 6, 5 could be suggested for planting to improve the air quality of the area.

Key words : Temperature, Carbon dioxide, urban agglomerations, green belt, suitable ability Index (SAI).

INTRODUCTION

Building activity provides a kind of habitat that gives rise to housing and shelter for the people. This built environment with the people results in the expansion of the city forming urban complexes and habitats. Building activity has grown both vertically and horizontally depending upon the availability of the area. Urbanization with increase in population have developed into slums. Urban agglomeration has also brought in industrial townships in and around the urban centres. An urban city or town disturbs the climate and creates a substantial source of pollution because of gathering of its dwellers who live, work and move with in and around the boundaries. The urban climate has a special influence on urban heat and on freshness, stability and dispersion of pollutants.

There are only a few studies available on effects of land use on local urban climate and the changes that have been observed in local climate (Oke,1971, 1991; Bitan, 1983; Landsberg, 1970, Sham, 1973) Boodhoo, 1991). There are hardly a few studies available with reference to Indian context (Sharma and Sainath, 1991; Subbaiah et al., 1991; Padmanabha-murthy, 1986). Further, Human comfort depends on temperature humidity, freshness of air, CO_2, SO_2, NO, dust and particulate matter in and around the built environments of a city. In order to reduce air pollution, improve air quality and provide cooling effect, the employment of plants in absorbing heat and pollutants is a very useful step in this direction. In the present study an attempt has been made to evaluate the thermal behaviour and the air quality of the urban Hyderabad *vis-a-vis* the role of certain plants in improving air quality by reducing dust and particulate matter, SO_2 and CO_2 from the inside and outside of the built environment.

METHODOLOGY

Air quality has been determined by analysing levels of sulphur dioxide in air by Katz's (1969) method. Carbon dioxide concentration by Wilson's method (1982) and dust fall deposition on plants by the method adapted by Sundaresan (1978) in NEERI's course mannual.

GEOGRAPHICAL LOCATION

Hyderabad is the fifth important city in South Central part of India. Hyderabad is located on 17° 22' N and 78° 27' E on the banks of the River Moosi with an elevation of 545 m above mean sea level. The city is characterized with large water bodies such as Osmansagar lake on the west. Himayatsagar lake and Mir Alam tank on the South West and Hussain sagar in between twin cities of Hyderabad and Secunderabad. Warm and cool areas characterized with high and low pollution are shown in the map of Hyderabad region (Map 4.1).

GENERAL CLIMATE

The climate of Hyderabad is generally hot and dry (Map 4.1) and is characterized with seasonal variations of winter from November to February, Summer from March to June and rainy season from July to October. The mean maximum temperature in summer ranges from 30-41°C, in winter is 27-31°C and in rainy season is 29-35°C. The range of mean minimum temperature in summer is 25-28°C in winter is 12-20°C and in rainy season is 20-28°C. On rare occasions the temperature reaches as high as 46°C and as low as 9°C. Wind speeds are less in the post monsoon months and cold season. The velocity of wind reaches its peak in the late after noons of the summer. The annual rainfall is mainly due to south-west monsoons which is almost 800 mm and occasionally the city receives rains of the nature of hail storms during the summer. The city experiences a dry humidity. The moisture contents of the atmosphere is generally high in July to September and low in May and in some winter months because of dry winter with severe cold.

GENERAL SOIL CONDITIONS AND TOPOGRAPHY

The city area consists of almost brackish brown soils with loamy moderately coarse texture. The city also consist of hillocks upto 15 meters and barren rocks were found exposed in many places.

GENERAL VEGETATION

The city vegetation consists of cultivated shrubs, trees and other garden plants for the panoromic beauty and avenue plantations. The natural vegetation at the outskirts of the city is represented by scrub jungle. A number of fruit orchards and vine yards and crop fields dot the landscape of the city outskirts.

POPULATION

The present population according to 1991 census is around 32,00,000 of urban Hyderabad. However, the population of rural and urban Hyderabad together is 42,80,261. There has been tremendous increase in the human population of Hyderabad city from 12 lakh in 1961 to 42 lakhs in 1991 which cover a period of 3 decades.

Table 4.1 : Population of Hyderabad

Year	1961	1971	1981	1991	
Population	12,54,403	16,18,484	22,40,508	32,00,000 (Urban City)	42,80,261 (Urban Rural)

Source : Andhra Pradesh Year Book, 1992

However, the land area at the disposal is the same as that was 3 decades ago. With the increase in population, there has been no increase in the facilities, resulting in closer aggregation of the population forming slums.

Table 4.2 : Total Number of Air Pollution Sources in Hyderabad

1. Saw mill, Plywood, Wood furniture.
2. Stone quarrying
3. Pottery-chinia, earthenware
4. Structural clay and Brick works
5. Plastic and Vinyl products
6. Glass industries
7. Cement plant
8. Cement-concrete products
9. Iron-Steel smelters. Works and foundries
10. Basic Chemical Fertilizers
11. Chemical related products
12. Pharmaceuticals
13. Rubber-latex products
14. Smoke producing houses
15. Palm oil mills
16. Palm, Vegetable oil Products, and refineries.
17. Petroleum products.
18. Fish, Crustaceae and similar Food
19. Animal Feeds.
20. Agricultural wastes etc.

Table 4.3 : Hyderabad Circles with Number of Slums

Circle	*Area covered*	*No. of Slums*
I.	Erakulabasti, APHB Colony, Dilsukhnagar, Gachibowli, Kumarwadi, Musarambagh, Phoolbagh, Saidabad, Charminar.	80
II.	Vattapally, Zoogate, Chidibazar, City College, Gundipet.	57
III.	Adikmet, Vaddrabasti, Parsigutta, Old MLA quarters, Bholakpur, Fever Hospital, Nagamiah Kunta, Kachiguda X road, Golnaka, Indira Nagar, Zinda Tilismath Nagar,	101
IV.	Afzal sagar, Dhobighat, Hakeempet, Kumarwadi, Syednagar, Ziagauda.	56
V.	Ameerpet, Banjara Hills, SR Nagar, Yoususfguda.	66
VI.	Aghapura, Darulsalam, Chandan Bagh, Phoolbhag, Rd. Thopkhana.	32
VII.	Addagutta, Begumpet, Chilukalaguda, Mailargaddaa, Medibai, Warasiguda.	75

SELECTION OF THE STUDY SITE

Two types of areas have been selected namely low polluted and high polluted areas (Map 4.1). Low polluted areas comprise of permanent water bodies, parks, landscape gardens, farmlands. High polluted areas comprise of the places associated with Industrial structures, multistoreyed apartments and heavy automobile traffic.

Meteorological data has been obtained from India Meteorology Department. Air Port, Hyderabad and Urban data for the city area has been collected by installing a weather station in the city area for assessing the climate and the air quality of urban Hyderabad.

PLANT SELECTION CRITERIA

Following plants (Gamble, 1967) have been assessed as to how the levels of pollutants from the vicinity of different urban situations incorporated in the area on the basis of suitable ability index work in the

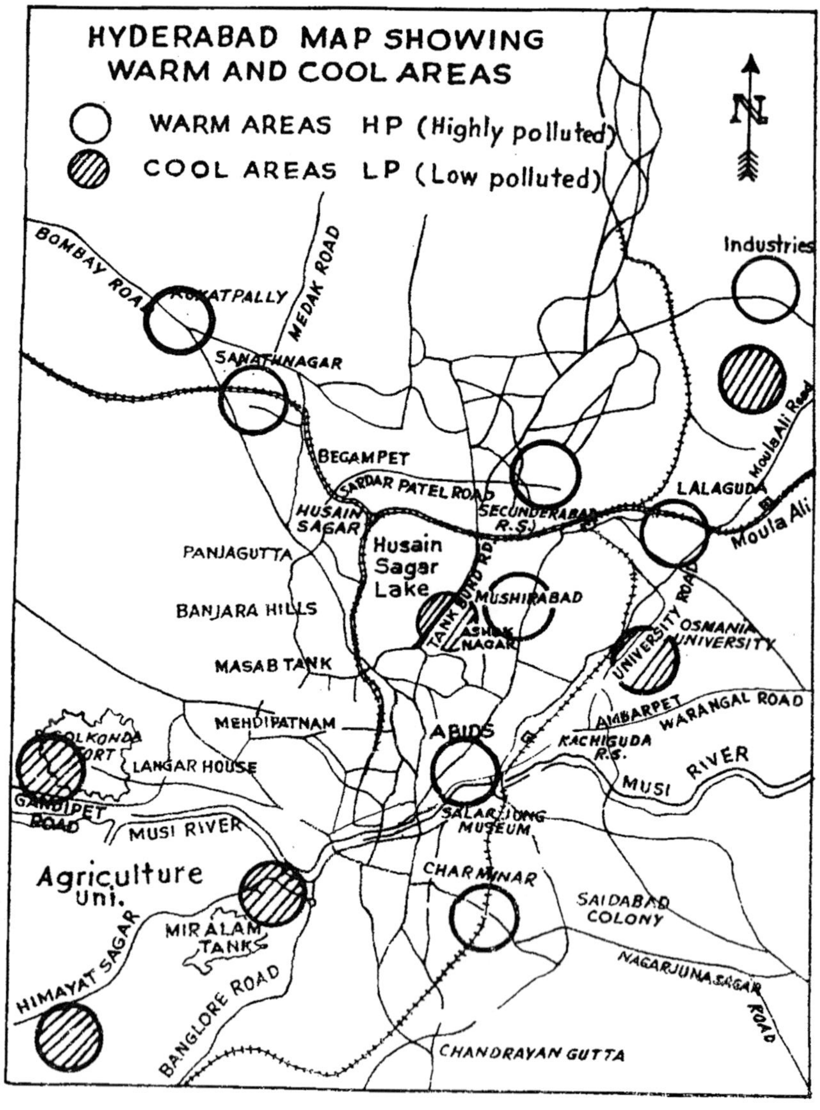
HYDERABAD MAP SHOWING
WARM AND COOL AREAS
WARM AREAS HP (Highly polluted)
COOL AREAS LP (Low polluted)
N
BOMBAY ROAD
KUKATPALLY
MEDAK ROAD
Industries
SANATHNAGAR
BEGAMPET
SARDAR PATEL ROAD
HUSAIN SAGAR
SECUNDERABAD R.S.
LALAGUDA
Moula Ali Road
Moula Ali
PANJAGUTTA
Husain Sagar Lake
BANJARA HILLS
TANK BUND RD
MUSHIRABAD
ASHOK NAGAR
UNIVERSITY ROAD
OSMANIA UNIVERSITY
MASAB TANK
MEHDIPATNAM
GOLKONDA FORT
LANGAR HOUSE
ABIDS
AMBARPET
WARANGAL ROAD
KACHIGUDA R.S.
MUSI RIVER
GANDIPET ROAD
MUSI RIVER
SALAR JUNG MUSEUM
Agriculture Uni.
CHARMINAR
SAIDABAD COLONY
MIRALAM TANK
HIMAYAT SAGAR
BANGLORE ROAD
NAGARJUNASAGAR ROAD
CHANDRAYAN GUTTA

Map 4.1

improvement of the environment. The method of Tiwari (1991) has been modified and used in developing a grading pattern of suitable Ability Index (SAI) Scale. The data is presented in the descending order of the index (Table 4.6).

Table 4.4 : The Existing Land use in Hyderabad

Area	*Area sq. km.*	*Land use Category*
Hyderabad + Secunderabad	199.66	Residential + Commercial + Industrial
Secunderabad contonement	40.17	Residential + Commercial + Open land
Uppal	8.39	Industrial + Agricultural
Ramanthapur	3.41	Educational + Telecommunication
Saroornagar	0.57	Residential
Moulali	3.00	Industrial
Lalaguda	1.58	Industrial + Railways
Malkaigiri	17.30	Residential
Qutabullapur	7.41	Residential
Fatehnagar	2.17	Industiral
Balanagar + Sanathnagar	29.53	Industrial
Kapra	11.75	Residential
Kukatpalli	16.62	Industrial + Agricultural + Open land
Moosapet	7.12	Residential
Bowenpalle	3.58	Residential + Agricultural + Open land
O.U. + Tarnaka	4.00	Educational + Residential

TRAFFIC PROBLEM

The high traffic volume in Hyderabad city is accompanied by problems of traffic congestion, parking spaces and automobile exhausts resulting in air pollution. Though there is no continuous recorded data of air pollution caused by traffic emissions, observation and experience have proved that the concentration of photochemical smog with carbon mono and dioxide, especially during the day is enough to identify the pollution hazards. The longer duration of any such situation results in creating discomfort to the people living in the area suffering from respiratory diseases etc.

RESULTS AND DISCUSSION

Fig. 4.1 indicates ombrotherm depicting the regional climate of Hyderabad. The characteristic feature of the Hyderabad climate is that there are almost 7 months of dry period running with the problem of water deficiency. Approximately 5 months of relatively moist to wet conditions with almost 814 mm of rainfall and 25.8°C temperature exist exhibiting tropical to sub-tropical character with deciduous vegetation and scrubs.

Fig. 4.2 shows a comparison between two sites one representing urban-industrial area and the other green belt area. The difference in the mean monthly maximum and minimum temperatures have been observed as shown in the curve. There is almost 2-3° C difference between these sites. Urban industrial area with higher values than the green belt area.

Table 4.5 shows temperature variations in 24 hours in two different urban conditions during winter season. It is observed that the factory sites, building and constructed areas are 7-11°C higher than the areas covered with thick vegetation and green belt conditions between 12-16 hours in the course of day (7-11°C more near the factory sites buildings and constructed areas than the areas with vegetation covered between mid day till evening) Nights are cool with a difference of almost 1°C.

The critical evaluation of Fig. 4.3 indicates that the urban-industrial site is characterized with the higher levels of sulphur dioxide and dust while the green belt area is encountered with lower levels. Fig. 4.4 also represents higher levels of carbon dioxide in the industrial site than in the green belt area. However, the general range of carbon dioxide varies from 250-600 ppm in different environments of urban Hyderabad agglomerations.

The present situation is the climatic condition that have been described in the text representing the change in the mean temperatures, and with a possible shift in occurrence of rainfall. This has been attributed to the uprising of the large number of industries (Table 4.2), in the vicinity of Hyderabad, increase in population (Table 4.1) and the slum areas (Table 4.3). They have resulted in the depletion of the water table, impairing the crop schedule of the rained crops. The information obtained from Table 4.5 indicates that the temperature is 7-11°C less in the green belt areas. The work of Sakakibara (1989), Yamashita (1992). Segal et al. (1988) also represents similar behaviour of the urban environment. Jauregui (1991) has shown how vegetation and artificial water bodies have favourably modified the climate of North East Mexico City-Texcoco.

Tropical — Dry climate with 7 dry months with water deficit period and the rainfall distributed over a period of ≈ 5 months with relatively moist conditions with deciduous vegetation and scrubs.

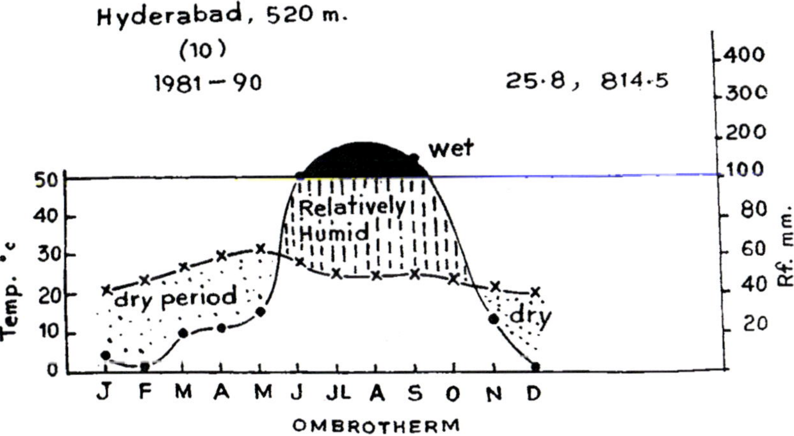

Fig. 4.1

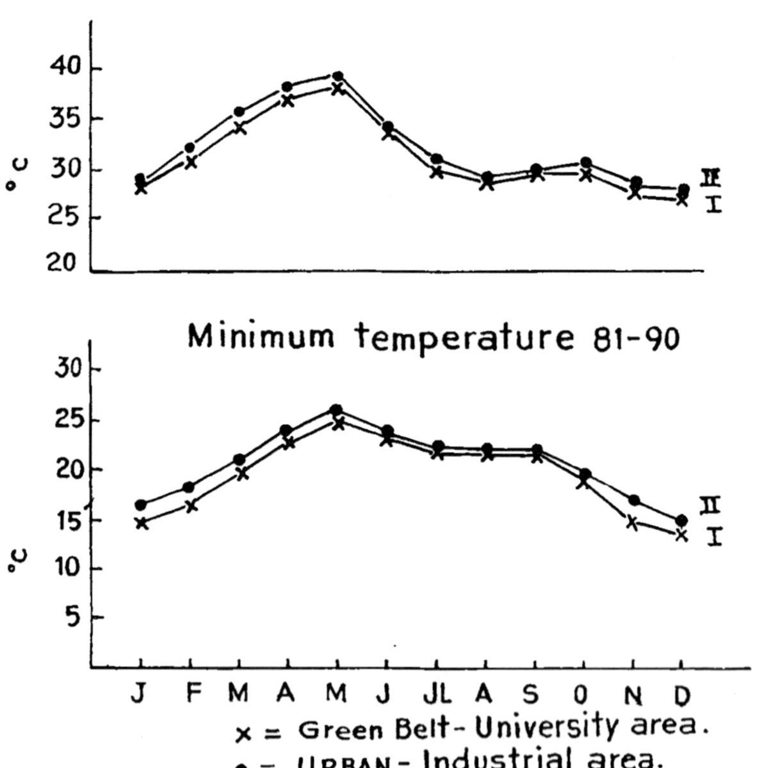
VARIATIONS IN TEMPERATURE BETWEEN TWO SITES
Maximum temperature 81-90
°C
40
35
30
25
20
II
I
Minimum temperature 81-90
°C
30
25
20
15
10
5
II
I
J F M A M J JL A S 0 N D
x = Green Belt - University area.
• = URBAN - Industrial area.

Fig. 4.2

Table 4.5 : Temperature variation in 24 hours in two different urban conditions (during winter season)

Condition	*6*	*8*	*10*	*12*	*2*	*4*	*5*	*8*	*10*	*12*	*2*	*4*	*6*
1. Concrete Jungle Building & Factories/Highly polluted site	20	22.5	29	32	34	37	26	24	22	22	21	20.5	20
2. Vegetation Jungle Vegetation Low polluted sites	22	22.4	24	25	26	26	27	25	23	23	20	19.0	21
3. Difference of temperature between Vegetation-Building area	-2	+0.3	+5	+7	+8	+11	-1	-1	-1	-1	+1	+1.5	-1

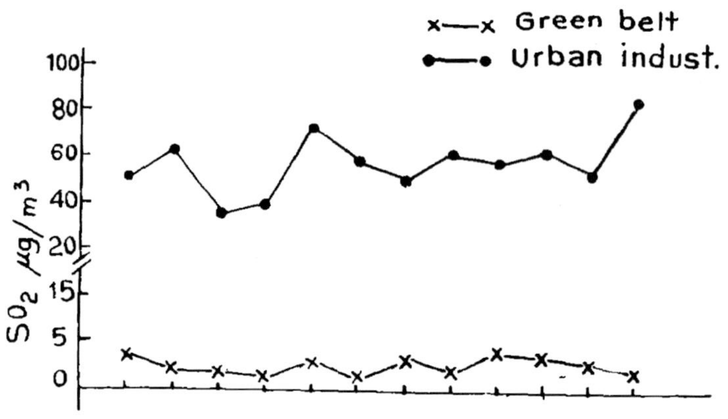
Variations in SO_2 in two sites
x—x Green belt
•—• Urban indust.
SO_2 $\mu g/m^3$
100
80
60
40
20
15
5
0

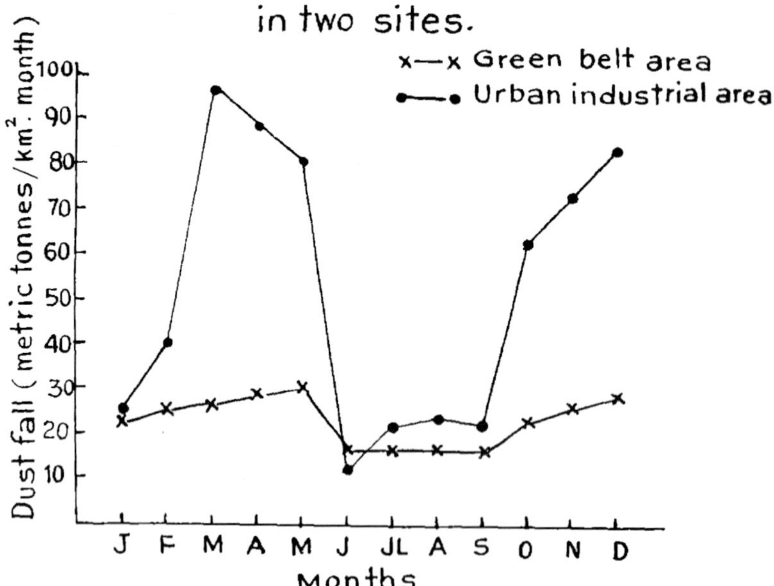
Variations in dust (MT/km²/month) in two sites.
x—x Green belt area
•—• Urban industrial area
Dust fall (metric tonnes/km². month)
100
90
80
70
60
50
40
30
20
10
J F M A M J JL A S O N D
Months

Fig. 4.3

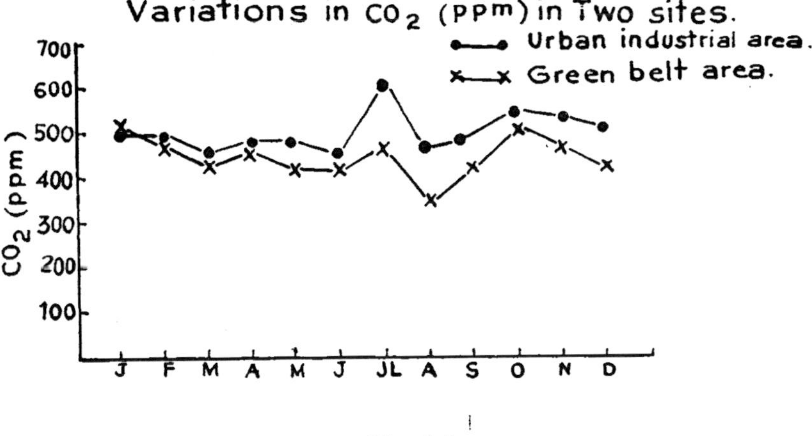
Variations in CO_2 (ppm) in Two sites.
Urban industrial area.
Green belt area.
CO_2 (ppm)
700
600
500
400
300
200
100
J
F
M
A
M
J
JL
A
S
O
N
D

Fig. 4.4

Table 4.6 : Arrangement of Plants as per Suitable Ability Index Pattern

S. No.	Name of the Plant	Total Plus	% Scoring	SAI Grade
1.	*Mangifera indica*	15	94	7
2.	*Artocarpus integrifolia*	13	81	6
3.	*Ficus benghalensis*	13	81	6
4.	*Ficus infectoria*	13	81	6
5.	*Mimusops elengi*	12	81	6
6.	*Dalbergia sissoo*	12	75	5
7.	*Shizigium (Eugenia jambolana)*	12	75	5
8.	*Ailanthus excelsa*	11	69	4
9.	*Ficus religiosa*	11	69	4
10.	*Saraca indica*	11	69	4
11.	*Acacia leucocephala*	10	62	4
12.	*Anthocephalus cadamba*	10	62	4
13.	*Butea monosperma*	10	62	4
14.	*Cassia siamea*	10	62	4
15.	*Ficus glomerata*	10	62	4
16.	*Madhuka latifolia*	10	62	4
17.	*Melia azadirach*	10	62	4
18.	*Putranjiva roxburghii*	10	62	4
19.	*Acacia nilotica*	08	50	2
20.	*Aegle marmeolus*	08	50	2
21.	*Bauhinia variegata*	08	50	2
22.	*Bougainvillea spectabilis*	08	50	2
23.	*Cassis fistula*	08	50	2
24.	*Cassia renigera*	08	50	2
25.	*Cassia merginata*	08	50	2
26.	*Cordea sebestena*	08	50	2
27.	*Erythrina indica*	08	50	2
28.	*Ficus carica*	08	50	2
29.	*Lagerstroemia speciosa*	08	50	2
30.	*Millingtonia hortensis*	08	50	2
31.	*Pongamia pinnata*	08	50	2
32.	*Tabebnia pentaphylla*	08	50	2
33.	*Albizzia lebbeck*	07	44	2
34.	*Casuarina equisetifolia*	07	44	2
35.	*Parkia biglandulosa*	07	44	3

S. No. Name of the Plant	Total Plus	% Scoring	SAI Grade
36. *Acacia auriculiformis*	09	56	3
37. *Enterilobium saman*	09	56	3
38. *Erythrina variegata*	09	56	3
39. *Grevillea robusta*	09	56	3
40. *Morninga pterigospermum*	09	56	3
41. *Peltophorum plerocaarpus*	09	56	3
42. *Pinus longifolia*	09	56	3
43. *Spathodea campanulata*	09	56	3
44. *Terminalia arjuna*	09	56	3
45. *Feronia elephatnum*	06	37	1
46. *Grevillea pteridophylla*	06	37	1
47. *Hibiscus rosa sinensis*	06	37	1
48. *Phyllanthes embelica*	06	31	1
49. *Poinciana regia*	05	37	1
50. *Jacaranda mimosifolia*	06	31	1
51. *Parkinsonia aculeata*	05	31	1
52. *Santalum album*	06	31	1
53. *Sesbanea grandiflora*	05	.37	1

Further, the improve the climatic conditions existing in the area, certain species which fit into the grading pattern with respect to their performance as per suitable ability index (SAI) (Table 4.6) are suggested for plantations. More than 50 plants which can withstand to the changes in the climatic conditions, tolerate pollution level and have a good canopy structure etc. can be suitably used in the order of higher index value thus accepting the grading pattern of Tiwari (1991). Earlier Raza et al. (1991) have reported the cooling effect and improvement of climate in the certain areas of Hyderabad by planting certain trees. Fritz (1991) has also reported the effect of vegetation on the urban topoclimate and microclimate in German conditions. The climate - vegetation relationship probably may go a long way in improving the urban areas *vis-a-vis* the climate of the urban environment and beautification of the urban agglomerations with proper land scaping patterns.

CONCLUSION

The temperature rise in the urban Hyderabad over a small unit space and time has been recorded as a consequence of rise in population, industrial emissions and the rise in vehicular traffic. Greening of the areas with plantations (Green belt programmes) have considerably reduced temperature (7-11°C) by producing cooking effect and improving the air quality. Grading the suitable ability index has helped in selection of plants for developing green belts.

ACKNOWLEDGEMENTS

The author thanks the Departments of Botany and the University authorities for necessary help and encouragement during these studies.

REFERENCES

1. Avissar, R. 1992. Potential effects vegetation on the thermal environment of urban areas. In (Eds). T. Katayama and J. Tsutsumi, 2nd Intl, Symp. Conf, Urban Thermal Environment, special in Tohwa, pp. 103-104, Tohwa Institute for Science. Tohwa University, Fukuoka. Japan.

2. Boodhoo, Y. 1991 Living in the tropics. Can it be better. *Energy and Buildings 16*. 817-822.

3. Bitan, A. 1983. Applied climatogy and its contribution to planning and building - the research experience. *Habitat Intl. 7*. 125-145.

4. Fritz Wilmers 1991. Effect of vegetation on urban climate and buildings. *Energy and Buildings 15*. 507-514.

5. Gamble, J.S. 1967. *Flora of Presidency of Madras.* (Vols. 1-3). Botanical Survey of India, Calcutta.

6. Jauregui, E. 1991. Effects of Revegetaion and new artificial water bodies on the climate of North east Mexico City. *Energy and Buildings* 15 : 447-455

7. Katz, M. 1969. *Measurements of air pollutants*. Guide for the *selection of Methods* pp. 149-155. W.H.O. Geneva.

8. Landsberg, H.L. 1970. Meteorological observations in Urban areas *Meteorol: Monogr. 33*. 91-99.

9. Oke, T.R. 1979 *Review of Urban Climatology*, 1974-1978. Tech. Note 169 pp. 1-100. WMO Geneva.

10. Oke, T.R. 1986, *Urban climatology and its application with special regard to Tropical areas.* Publ. No. 652 pp. 1-25 WHO Geneva.

11. Oke, T.R. 1991. *Bibliography of Urban Climate.* 1981-1988. WHO. WCP Series. WHO Geneva.

12. Padmanabhamurthy, B. 1986. Some aspects of the urban climates of India. Proc, Conf. Urban Climatology and its application with special regard to Tropical areas. Publ. No. 652. 136-165 WMO, Geneva.

13. Raza, S.H. Murthy, M.S.R., Bhagyalakshmi. O. and Shylaja, G. 1991. Effect of vegetation on urban climate and healthy urban colonies. *Energy and Buildings,* 15. 487-491.

14. Sakakibara, Y. 1989. The rise in temperature on railway station platforms. *Geographical Review of Japan. 62A-4.* 311-319.

15. Segal, M., Avissar, R., Mc Cumber. M. and Pielke. R.A. 1988. Evaluation of vegetation effects on the generation and modification of mesoscale circulations. *J. Atmos. Sci... 45.* 2268-2292.

16. Sham, S. 1976. Current state of the environment within and around Kuala Lumpur-Petaling. Jaya. A review *Akademlika.* 10. 37. 40.

17. Sham, S. 1973, Observations on the effect of a city's form and function on temperature patterns. A case of Kualalumpur *J. Trop. Geogr.* 36 : 60-65.

18. Sham, S. 1973. The urban heat island : its concept and application to Kuala Lumpur. *Sains. Malaysiana.* 2 : 53-64.

19. Sharma, A.A.L.N. and Sainath, B.V.H.N., 1991. Studies on urban climatic variations. *Energy and Building* 15. 119-128.

20. Subbaiah, S., Vasantha, V. and Kaveri Devi, S. 1991. Urban climate in Tamil Nadu. India : A statistical analysis of increasing urbanization and changing trends of temperature and rainfall. *Energy and Buildings.* 15, 231-243.

21. Sundaresan, B.B. 1978. *Air quality monotoring A. course Manual.* National Environmental Engineering Research Institute. (NEERI) Nehru Marg. Nagpur.

22. Tiwari, S.L. 1991. Studies of Air Pollution Tolerance Indices in some planted trees in urban areas of Bhopal with reference to Ecoplanning of industrial areas. Ph. D. Thesis, Barkatullah University, Bopal.

23. Wilson, J.F. 1982. Energy balance and carbon outflow in a grassland ecosystem. I (Eds.) Whittakar. J.B., Davies. W.J., Proceedings of Ecological Process. pp. 153-163. Lancaster. P.A. University of Lancaster, (Pennesylvania), P.A.

24. Yamashita, S. 1992. The detailed structure of heat island phenomena estimated by the electric car in Metroplitan Tokyo, In : 2nd Symp. Conf. on Urban Thermal Environment. Special in Tohwa 101-102. Tohwa Institute for science. Tohwa University, Fukuoka, Japan.

THE INDOOR ENVIRONMENT AND ITS IMPACT IN HUMAN HEALTH

D.M. Tripathi

INTRODUCTION

Increasing attention is being directed to the indoor environment as a focus of exposure to air pollutants. Specific pollutants may be more concentrated indoors than outdoors, and the indoor environment can provide a unique exposure situation for more pollutants. Pollutants released by microorganisms have yet to receive intensive unified study and remain poorly known. Microbial pollutants include vegetative microbial cells, their reproductive units and metabolites. Most microbial pollutants can probably act as sensitizing agents in susceptible hosts, whereas some cause infectious diseases and other produce toxins with both acute and chronic health effects.

ECOLOGY OF MICROBIAL POLLUTANTS

Although many microorganisms can survive on environmental surfaces for varying lengths of time, they only very rarely cause diseases when they re-entain into the air. Most diseases caused by parasitic micro-organisms are acquired by direct contact. Many microorganisms can cause human diseases when they are airborne in sufficient numbers. For example, pollen grains, fungal spores (moulds), insects, mites and their excreta are major pollutants that lurk indoor causing chronic rhinitis and excerbating asthma in unsuspected victims :-

A. *Pollen and Spores* : Intact pollen grains and fungal spores are presumed to be the primary carriers of allergens. The outdoor air is usually well supplied with pollen and fungal spores. These pollens and fungal spores freely penetrate interiors through open windows, mechanical air intakes. Therefore, no interior environment is completely free of pollen and fungal spores and concentrations in normally ventilated (non-airconditioned interiors usually are directly correlated with concentrations of outdoor air. In addition, many spores are present per gram surface dust in most enclosed spaces. Tripathi and Parikh (1983a) detected presence of pollen and fungal spores in house dust samples and pointed out that though the amount of pollen grains and fungal spores in house dust sample is minute as compared to other formites, yet it is still possible that these tiny amounts of pollen grains and fungal spores may play role in precipitating symptoms of rhinitis or bronchial asthma in sensitive individuals.

Many microorganisms also produce volatile organic compounds that can be mucosal irritants or possibly systemic toxins. The moldy or mildewy odor associated with mold contamination in indoor environments is the result of low levels of these volatile compounds in the air (Burge, 1990). Health effects of these volatile compounds has not yet been studied, but they may well contribute to complaints of headache, eye and throat irritations, nousea, dizziness and fatigue in subjects occupying contaminated interiors.

B. *Dust and dust mite* : House dust is main source of indoor allergens. House dust contains many important indoor allergens such as animal dandars, fungal spores, pollen grains, cockraches, and house dust mites. Allergic asthma, rhinitis or conjunctivitis diseases caused by airborne allergens have been associated with inhalation of house dust (Caldas & Lockey, 1990). Though, house dust contains variety of allergens-organic and inorganic materials such as fibers, mamalian dandars mite and mite feces, mold spores and pollen grains; probably, the most important indoor allergens are mites; the small archanids or insect like creature that live in house dust. They can not be seen without the aid of a microscope, but they are present in large numbers in house dust. Dust mite allergy is recognised as an important clinical problem in many areas of the world. The World Health Organisation (W.H.O.) has recognised mites as the universal health problem. The dust mites can be found on floors, in carpets, mattresses, over stuffed furniture, bedding and droperies. Optomal mite growth occurs at a relative humidity of 75% and temperature between 22°C and 26°C. Mites can not survive where the relative humidity is less than 50%. Dust mite osmoregulate through their cuticle and that is why require high ambient humidity to prevent water loss.

The mite's feces have been identified as major allergen (Anderson & Ownby 1990). There are three different ways mite may secrete or excrete products :- egg laying, lateral 'Oil' gland secretion and feces production (including guanine secretion). Currently, there is no evidence that eggs contribute allergens to house dust. Although, Secretions of lateral glands have not been demonstrated to be source of allergen, their chemistry needs further study. One of the major known allergen is protease, probably related to digestion. Guanine is an end product of purine digestion (which can be used as specific marker of mite infestation).

C. *Mamalian dandars and proteins* : Allergens of house pets are potent sensitizers. The main sources of animal allergen are saliva and urine which dry on fur or pelt and are later aerosolized. Large amount of these allergens remain airborne for long time periods and are sufficiently small in size (Less than 1 Micron) Cat allergens are very potent sensitizers. The major allergen in cat extract is found in large quantities in cat pelt, saliva and subaceous glands of cat skins. Cat allergens were detected in class room dust samples collected from the floors (Munir et al. 1993).

Dog allergen is an important perennial indoor allergen. Allergenic sensivity to dog is not as common as other mammals, however, 5-30% allergic individuals have positive skin tests to doff extracts (Caldas & Lockey 1990). Dog albumin and dog epidermal antigens have been identified as major allergen in Poodle and Alsation epidermal extracts. Dog allergens are present in urine, epidermal scale serum and saliva.

D. *Insects and their debries* : House dust may be a common source of insect allergies. Insect species from at least 12 orders have been implicated in the development of respiratory allergies. It is evident that exposure to insects in genetically predisposed individuals may result in sensitization to their allergens. Stinging and biting arthopods are important initiator of allergic disorders. Cockroaches can cause inhalent allergies as well as subcutaneous sensitivity and allergic individuals have a higher prevalence of positive skin test reactions to cockroach extracts than do non-allergic subjects. Several studies have documented an IgE mediated allergic response to cockroach allergens among urban patients with asthma and described the presence of immediate and late-phase reaction to bronchial challenges with cockroach allergen (Kang 1990).

ENVIRONMENTAL CONTROLS

Environmental controls can be effective, but to make it worthwhile, it must acctually reduce allergen exposure. Because we spend much time in indoor, maximum importance to be given to reduce indoor allergen exposure. Patients sensitive to house dust are advised to clean their bed rooms to reduce exposure to dust. The most successful and direct treatment of house dust mite allergic asthma is the removal of any identificable offending allergens. Although, the beneficial effect of such avoidance measures have long been proclaimed, a few control trials of usefulness have been reported :-

1. *Avoidance measures for pollen and fungal spores* :- Air cleaning/ filtration can control airborne indoor allergens. HEPA filter is the most efficient mechanical filter and is efficient to remove both larger and smaller particles. Air conditioning also helps to keep indoor air clean by filtering outdoor air borne pollen and fungal spores. Mite counts and mold spores are also reduced by air conditioning because indoor humidity is reduced. The air filtration device has been shown to be very effective in lowering indoor pollen and spores densities.

Fungal colonisation and fungal spore densities within the home can be decreased by ulterning the conditions under which fungi thrive. Thus mold sensitive individuals can decrease their exposure by reducing the humidity in the home; decreasing the amount of organic material in indoor environment; using fungicide or house hold germicide.

2. *Avoidance measures for animal dandars* : Eviction of pets is the best remedy in case of house pet allergy. When individuals have serious allergic disease, especially asthma related to cat or dog allergy, the removal of an animal is an almost necessity. Removing the animal for a few hours or even days, or going away from whome for a week does not result in sufficient improvement to demonstrate that the animal is offender. Keeping pet may result in continued chronic inflammations in the lungs. In the long run this may lead to significant damage that could have been avoided by removing the animal.

3. *Avoidance measures for dust and mites* :- Effective control measures for mites should incorporate the removal of allegen pool of fecal material and dead mites and reduction or elimination of living population. The best way to avoid mite infestation is by removing as many objects that collect dust from the home as a possible, particularly in the bed room. Other measures to reduce mite concentration include removing of carpets and rugs. The important measures for reducing levels of mite allergens are as follows :-

1. Replace feather pillows with synthetic and wash regularly in hot water.

2. Bed linen should be regularly washed in hot water.

3. Mattresses and box spring should be encased in plastic.

4. Mattresses, base of bed, rugs, carpets and upholstered furniture should be vacuum regularly.

5. Use of airconditioners, dehumidifiers and air-filtration devices may be beneficial.

Results of studies as environmental control measures to reduce symptoms in mite allergic individuals are still contradictory (Caldas & Lockey 1990). Nevertheless, improvements in asthma symptoms and decreased bronchial hypher reactivity were observed in allergic patients after they were removed from mite infested environments. On the other hand, in a number of studies, patients have not shown improvement after intensive cleaning and encasing of mattresses and pillows in their environments. It may be that these patients had multiple allergic.

Mite allergens increase in proportion to mite concentrations. A threshold level of 100 miles or 0.6 (micro gram) guanine in each gram of house dust will trigger genetically prone individuals to produce mite specific IgE. These concentrations frequently occur indoors specially in humid regions. Concentration of more than 500 mites per gram of dust are a major risk factor for mite-allergic asthma patients. In Indian environment more than 4000 mites have been reported from humid regions (Tripathi and Parikh 1983b). Estimates of mite allergen concentrations can established whether patients are being exposed to large quantities of mites and mite allergens, and help to evaluate the effect of environmental controls. At present these evaluates are used for research purposes but they may eventually be as clinically useful as pollen counts, to evaluate potentially mite sensitive patients.

Effective control measures should incorporate the removal of the allergen pool of mite feces and dead mite and reduction of living mite population by use of acaricides or other measures such as removal of carpets etc.

Some investigators regard reduction in humidity as the primary method of controlling mite allergen and also imply that it is difficult to control mites without reducing humidity (Korsgaard 1982). In any area it is necessary to analyse the source of humidity because the relevant control measures are strikingly different in different climate and cultures. However, modern energy conserving building construction with reduced ventilation may be associated with increased indoor humidity and consequent mite growth. Indoor humidity can only be decreased by dehumidifiers or air-conditioning. However, these measures are not always feasible in developing countries. In many tropical countries, the design of houses is very different with very little carpeting, upholstered furniture or bedding, therefore there may not be suitable mite nests.

Without present understanding and technical progress it is possible to define protocols for reducing levels of indoor pollutants and the use of basic knowledge will be helpful towards development of an approach that will be useful for treating specific cases.

REFERENCES

1. Anderson J.A. and Ownby D.R. 1990. Indoor allergens and pollutants : what clinician need to know. *Current Issues in Allergy and Immunology* 1 (1) 3-14.

2. Burge. H. 1990. Bioaerosols : Prevalence and health effects in the indoor environments. *J. Allergy & Clinical Immune.* 86, 687-701.

3. Caldas E.F. and Lockey R.F. 1990. The importance of environmental changes to control mite allergens. *Current Issues in Allergy & Immunology* 1 (1) 18-21.

4. Kang B. 1990. Cockroach allergy. Clinical Rev. Allergy. 8, 87.

5. Korsgaard J. 1982. Preventive measures in house test allergy. *Am. Rev. Res. Dis.* 125, 80-84.

6. Munir A.K.M. *et al.* 1993. Allergens in school dust. *J. Allergy & Clinical Immunology,* 91 (5) 1067-1074.

7. Tripathi D.M. and Parikh K.M. 1983 a. Mite fauna and other allergens present in the house dust in Bombay. *Lung India* 1 (4) : 147-151.

8. Tripathi D.M. and Parikh K.M. 1983 b. Mite fauna in the house dust of asthmatic patients. *Medicine and surgery.* 23 (7) : 17-20.

EVALUATION OF TREE SPECIES FOR PLANTATION IN AN INDUSTRIAL COMPLEX

S. Tiwari and S. Bansal

ABSTRACT

The life support system on this planet consist of air, water, land and flora and fauna. Normally these are mutually interconnected and also interdependent. In an attempt to reach materialistic target, man has chalked out an ambitious plan of rapid industrialisation and urbanization. In this venture he has not only destroyed the plant cover built up meticulously by nature over millions of years, but has also polluted air, water and land, so much so that development has become synonymous with deforestation and desertification and progress with pollution. To avoid any damage to the environment, it is necessary that industrialisation and urbanization should be carefully planned. A new approach has evolved in recent years to grow green plants in and around industrial and urban areas.

INTRODUCTION

Air and water, being the prime of life, should be protected from the evil of pollution as depletion in their quality may lead to serious consequences including many human health hazards and an endangered environment for the live stock. Even possibility for depletion in the quality of genetic material cannot be ruled out. Inspite of this, a progressive country can avoid neither industrialization nor urbanisation. Economic growth is thus being attained at the cost of ecological damage. In order to curb the menace of air pollution, use of mechanical collectors, fabric collectors, wet scrubbers, electro-static precipitators and of fume incineration has often been proposed. Erecting tall stacks to facilitate air dispersal of emission is also recommended. But, as all these devices are of mechanical nature possibility of their occasional failures cannot be ruled out leading to disasters like toxic gas leakage at Bhopal in which thousands of human beings and animals lost their lives. Moreover, all these devices are expensive demanding a big budget which is not available. Ultimately some cheaper second grade and less efficient systems are assembled leading to an ineffective control of pollution. This makes the problem grave and demands an immediate remedial recourse.

A new approach has evolved in recent years to grow green plants in and around industrial and urban areas. Capacity of plants to reduce air pollution is very well known (Tingey, 1968; Spedding, 1969; Bennett and Hill, 1973). To check the spread of such air pollutants emitted from an industrial complex, it is recommended to grow a green vegetation around by many scientists (Fleming, 1967; Bernatsky, 1969; Warren, 1973; Agrawal *et al.*, 1988 and 1989, Kapoor and Gupta, 1989 and Tiwari, 1991).

It is being increasingly realised that simultaneous to the planning of an industrial complex, a parallel tree plantation programme should also be launched so that when the industry starts production and release of emissions in the air, the green vegetational belt grown around is ready to act as a vast sink for the pollutants.

Selection of tree species which can be grown around an industrial complex is a tedious task and requires a complete know how about plants and their expected behaviour in a polluted environment. It is known for certain that plants differ considerably in their responses towards pollutants (Rao, 1985); some are highly sensitive and show immediate symptoms while others are hardy and tolerant. They can withstand the stress of pollution aptly well. Such studies are very significant as the information furnished can be utilized in making an industrial complex pollution free and suitable for human habitation. Many health hazards, so common under such environments, can be minimised, if not eradicated. The objective of present treatise is to identify pollution tolerant species and to suggest an ecological model in the form of green belt around an industrial complex to mitigate pollution.

MATERIALS AND METHODS

Leaves were collected from the three canopies of the experimental plants. Samples were collected in three different seasons *viz.* summer, rainy and winter season. Samples were collected between 7 and 8 a.m. in the morning. 10 gms. of mature leaves were taken for the purpose of biochemical analysis. The concentration of total chlorophyll was determined using the formulae given by Duxbury and Yentsch (1956) and MacLachlan and Zalik (1963). Relative Water Contents (estimated as % moisture content on oven dry weight basis) was calculated by the method proposed by Weatherly (1950). Ascorbic acid was estimated by method of Dubey and Amritphale (1984). Air Pollution Tolerance Indices of tree species were calculated by following formula proposed by Singh and Rao (1983).

$$APTI = \frac{A(T+P)+R}{10}$$

Where, A = Ascorbic acid contents (mg/gm of dry weight)

T = Total Chlorophyll (mg/gm fresh weight)

pH = pH of leaf extract

R = Relative Water Content (%)

Plants were analysed at various phytosocio-economic as well as a few biochemical levels for certain vital parameters. They are subjected to a grading scale just to find out model plant that can be grown around the industrial complexes. Following grading pattern was followed:

S. No.	*Grading Character*	*Pattern of Assessment*	*Grade Alloted*
1.	APTI	$x \pm \sigma$	
		$\sigma <$ - 1	+
		σ = 1	++
		σ + 1	+++
		σ + 2	++++
		σ + 3	+++++
2.	Tree Habit	Small Tree	-
		Medium Tree	+
		Large tree	++
3.	Canopy Structure	Sparse/irregular/globular	-
		Spreading crown/open/semidense	+
		Spreading dense	++
4.	Type of tree	Deciduous	-
		Evergreen	+

S. No.	*Grading Character*	*Pattern of Assessment*	*Grade Alloted*
5.	Laminar characters		
	(a) Size	Small	-
		Medium	+
		Large	++
	(b) Texture	Smooth	-
		Coriaceous	+
	Hardiness	Delicate	-
		Hardy	+
	Economic value	Less than three uses	-
		Three or four uses	+
		Five or more uses	++

The number of maximum grades that can be scored by a plant was thus 16. The total of grades was used as an indicator of expected performance of the plant in an industrial area. Plants scoring more than 60 per cent points were considered as model plants. The "best plus plant" is also being proposed. The plants were assessed finally into following categories. An Expected Performance Index (EPI) of plants was evaluated along an arbitrary 0 - 7 scale.

Grades	*Scores in %*	*Assessment*	
0.	Upto 30%	Not recommended for plantation	
1.	31 - 40	Very poor	(VP)
2.	41 - 50	Poor	(P)
3.	51 - 60	Moderate	(M)
4.	61 - 70	Good	(G)
5.	71 - 80	Very Good	(VG)
6.	81 - 90	Excellent	(Ex.)
7.	91 - 100	Best Plus Plant	(BPP)

RESULTS

Out of the 53 plant species investigated in the present studies, 'ollowing can be ranked as suitable for the plantation in three oblong orbits 'ound the point source on the basis of their Expected Performance Index EPI) in an industrial area.

S. No.	Grade	EPI Value	Name of Species
1.	Best Plus Plant	7	*Mangifera indica*
2.	Excellent	6	*Mimusops elengi, Artocarpus integrifolious, Ficus benghalensis, Ficus infectoria*
3.	Very Good	5	*Dalbergia sissoo, Eugenia jambolana*
4.	Good	4	*Melia azadarach, Ailanthus excelsa, Butea monosperma, Cassia siamia, Saraca indica, Acacia leucophloea, Anthocephalus cadamba, Madhuca latifolia, Putranjiva roxburghii, Ficus religiosa, Ficus glomerata.*

Mangifera indica was found to be the best plus plant BPP which can be grown in an industrial area and is expected to perform well. It has a dense tree canopy, evergreen nature and a foliage which might afford protection pollutant stress. The economic and aesthetic value of the tree is very well known and it may be recommended for profuse plantation in first curtain. As it is a slow growing plant, it is advisable to mix it up with some remarkably fast growing species like *Moringa pterygosperma* in the beginning. This plant has good APTI value but its canopy structure is very sparse; besides it is a deciduous tree. *Mimusops elengi, Artocarpous integrifolious, Ficus benghalensis, Ficus infectoria* are all evergreen beautiful trees with many more important economic characters like being fruit bearing ornamental or of medicinal value. All these trees may be recommended for first curtain and a mixed plantation of these species with *Mangifera indica* may prove to be beneficial.

A second curtain with a few trees of moderate EPI value and a few with high EPI value need be raised to absorb pollutants emitted by industries. *Dalbergia sissoo* and *Eugenia jambolana* are two such trees which can be recommended for profuse plantation in the second curtain. Both of them are evergreen with large canopies and they have their own economic value. *Dalbergia sissoo* is an ornamental avenue plant. *Eugenia jambolana* is valuable for its fruits and their medicinal value. Plantation of few species of first curtain specially *Mangifera indica* as an interceptive species may enhance the pollution tolerating capacity of this curtain. But

Table 6.1 : Evaluation of a few planted tree species of Bhopal on the basis of their APTI value and some Phyto-Socio-Economic Characters

S. No.	Name of the Plant	APTI	Assessment Parameters							Grade Alloted		EPI Grade
			Tree habit	Canopy Structure	Type of Tree	Leminar		Econo-mic im-portance	Hardi-ness	Total plus	% Scor-ing	
						size	texture					
1.	*Hibiscus rosa-sinensis*	++	-	-	+	++	-	+	-	06	37	1
2.	*Melia azedarach*	++	+	+	+	+	+	++	+	10	62	4
3.	*Aegle marmelos*	++	+	+	-	+	+	+	+	08	50	2
4.	*Feronia elephantum*	+	+	+	-	+	-	+	+	06	37	1
5.	*Ailanthus excelsa*	++++	++	+	-	++	-	+	+	11	69	4
6.	*Mangifera indica*	+++++	++	+	+	++	+	++	+	15	94	7
7.	*Moringa pteryqosperma*	+++++	+	-	-	+	-	+	+	09	56	3
8.	*Butea monosperma*	+++	+	-	-	++	+	++	+	10	62	4
9.	*Dalbergia sissoo*	+++	++	++	+	+	+	+	+	12	75	5
10.	*Erythrina indica*	+++	+	-	-	++	-	++	+	08	50	2
11.	*Erythrina variegata*	+++	+	-	-	++	-	++	+	09	56	3
12.	*Pongamia pinnata*	+	+	+	+	+	+	+	+	08	50	2
13.	*Sesbania grandiflora*	+++	-	-	-	+	-	+	-	05	31	1

S. No.	Name of the Plant	APTI	Assessment Parameters									EPI Grade
			Tree habit	Canopy Structure	Type of Tree	Leminar		Econo-mic im-portance	Hardi-ness	Grade Alloted		
						size	texture			Total plus	% Scor-ing	
14.	*Bauhinia variegata*	+	+	+	-	++	-	++	+	08	50	2
15.	*Cassia fistula*	+++	+	+	-	+	-	+	+	08	50	2
16.	*Cassia renigera*	+++++	+	-	-	+	-	-	+	08	50	2
17.	*Cassia siamea*	+++	+	+	+	+	+	+	+	10	62	4
18.	*Cassia marginata*	+++	+	+	+	+	-	+	+	08	50	2
19.	*Parkinsonia aculeata*	+++	+	-	-	-	-	-	+	05	31	1
20.	*Peltophorum ferrugineum*	++	++	++	+	+	-	-	+	09	56	3
21.	*Poinciana regia*	+	++	+	-	+	-	-	+	06	37	1
22.	*Saraca indica*	++	+	++	+	++	+	+	+	11	69	4
23.	*Acacia nilotica*	++	+	+	+	-	-	++	+	08	50	2
24.	*Acacia leucophloea*	++++	+	+	-	+	-	++	+	10	62	4
25.	*Acacia auriculiformis*	+++	++	-	+	+	+	-	+	09	56	3
26.	*Albizzia lebbeck*	++	++	+	-	-	-	+	+	07	44	2
27.	*Enterolobium saman*	+	++	+	+	+	+	+	+	09	56	3
28.	*Parkia biglandulosa*	++++	++	-	-	-	-	-	+	07	44	2

S. No.	Name of the Plant	APTI	Assessment Parameters							Grade Alloted		EPI Grade
			Tree habit	Canopy Structure	Type of Tree	Leminar size	Leminar texture	Econo-mic im-portance	Hardi-ness	Total plus	% Scor-ing	
29.	*Eugenia jambolana*	++	++	++	+	++	+	+	+	12	75	5
30.	*Laqerstroemia speciosa*	+	+	+	-	++	+	+	+	08	50	2
31.	*Terminalia arjuna*	+	++	+	-	++	+	+	+	09	56	3
32.	*Anthocephalus cadamba*	++	++	+	-	++	+	+	+	10	62	4
33.	*Madhuca latifolia*	+++	++	++	-	++	-	-	+	10	62	4
34.	*Mimusops elengi*	+++	++	++	+	++	+	+	+	13	81	6
35.	*Cardia sebestina*	+++	-	+	-	++	+	-	+	08	50	2
36.	*Jacaranda mimosifolia*	+	+	+	-	-	-	+	+	05	31	1
37.	*Millingtonia hortensis*	++	++	-	+	+	-	+	+	08	50	2
38.	*Spathodea campanulata*	+++	++	-	-	++	+	-	+	09	56	3
39.	*Tabebuia pentaphylla*	+++	++	-	-	++	-	-	+	08	50	2
40.	*Bougainvillea spectabilis*	+++	+	-	+	+	+	-	+	08	50	2
41.	*Grevillea robusta*	++	++	-	+	+	+	+	+	09	56	3
42.	*Grevillea pteridifolia*	+	+	-	+	+	+	-	+	06	37	1
43.	*Santalum album*	++	+	-	+	+	-	-	+	06	37	1

S. No.	Name of the Plant	APTI	Assessment Parameters							Grade Alloted		EPI Grade
			Tree habit	Canopy Structure	Type of Tree	Leminar		Econo-mic im-portance	Hardi-ness	Total plus	% Scor-ing	
						size	texture					
44.	*Phyllanthus emblica*	+	+	+	-	-	-	+	+	05	31	1
45.	*Putranjiva roxburghii*	+++	+	+	+	+	+	+	+	10	62	4
46.	*Artocarpus integrifolius*	+++	++	++	+	++	–	+	+	13	81	6
47.	*Ficus religiosa*	+++	++	+	-	++	+	+	+	11	68	4
48.	*Ficus benghalensis*	+++	++	+	+	++	+	++	+	13	81	6
49.	*Ficus glomerata*	+	++	+	-	++	+	++	+	10	62	4
50.	*Ficus carica*	++	+	+	-	++	+	-	+	08	50	2
51.	*Ficus infectoria*	+++	++	+	+	++	+	++	+	13	81	6
52.	*Casurina equisetifolia*	++	++	-	+	-	-	+	+	07	44	2
53.	*Pinus longifolia*	+++	++	-	+	-	+	+	+	09	56	3

the curtain may have *Dalbergia sissoo* and *Euginia jambolana* as the dominant species.

The third curtain trees may be with moderate APTI but with good aesthetic value. In this curtain too, interception of a few trees of I and II curtain may increase the pollution, reducing capacity of the curtain. *Melia azedarach, Ailanthus excelsa, Butea monosperma, Cassia siamia, Saraca indica, Acacia leucophloea, Anthocephalus cadamba, Madhuca latifolia, Putranjiva roxburghii, Ficus religiosa* and *Ficus glomerata* are the trees with moderate EPI value and are assessed as good for plantation. Almost all these species are ornamental avenue trees and are grown in different parts of the city along road side.

Outside all these curtains a thick buffer zone curtain of mixed species of all three curtains be developed. In this curtain a large scale plantation of *Casurina equisetifolia* may prove to be highly useful. This plant species besides being hardy and ornamental, is known for its dust catching capacity (Katyal and Satake, 1989). It is fast growing and makes suitable wind breaks, hedge row, groves and wood lots. Its fire wood is valued and the mycorrhizal roots are known to increase fertility of the soil. *Acacia auriculiformis* is another handsome tree which has been grown with success in dry, poor and murramy soils in several parts of the country.

Some species have a low APTI value and are valuable for an eco-planner. They may act as bio-indicators of pollution. Such species should be planted in all the belts to monitor the air pollution level in the ambient air. A gradient of pollution from point source to outwards can be drawn on the basis of the symptoms expressed by these species. *Feronia elephantum, Pongamia pinnata, Bauhinia variegata, Poinciana regia, Enterolobium saman, Lagerstroemia speciosa, Jacaranda mimosifolia, Grevillea pteridophylla, Phyllanthus emblica* and *Ficus glomerata* were found to have very low APTI and may serve the purpose of bio-indicators. These species should be planted in all the vegetational belts in an interceptive manner.

EXPLANATION TO ECO-PLANNING MODEL

Plantation in the direction of the wind

Trees recommended for plantation in different zones:

Zone I *Mangifera indica, Moringa pterygosperma, Mimusops elengi, Ficus religiosa*

Zone II : *Dalbergia sissoo, Eugenia jambolana* and trees of first zone as interceptive species

Zone III *Melia azedarach, Ailanthus excelsa, Butea monosperma, Cassia siamea, Saraca indica, Acacia leucophloea, Anthocephalus cadamba, Madhuca latifolia, Putranjiva roxburghii, Ficus glomerata* and trees of first and second zones as interceptive species.

Buffer Zone : *Casurina equisetifolia, Acacia auriculiformis, Cassia renigera, Parkia biglandulosa* and mixed trees of all the three zones.

Following sensitive species to be grown in all the 4 zones as interceptive species:

Bauhinia variegata, Feronia elephantum, Sesbania grandiflora and *Poinciana regia.*

DISCUSSION

The plantation recommended above is not for any specific industry. The air pollution tolerance index of the few planted tree species has been evaluated giving weightage to some other relevant characters. An attempt has been made to grade a plant with reference to its potentiality to grow in an industrial complex. The performance of a plant under such environment is visualized only hypothetically looking at the physiological and phyto-economical characters of the plant. That is why it is coined as an Expected Performance. In the ecoplanning proposed by Rao Bhilai Steel Plant and its adjoining areas, it has been recommended that plants be grown in oblong orbits from point source of affluent emission. The selection of the plant was made on the basis of their air pollution tolerance level. structural geometry and phenology besides keeping in mind the bioaesthetic and ornamental value of the plant species selected for plantation proposed. Agrawal and his associates (1988-89) have proposed plantation in and around two cement factories situated at Mandhar and Akaltara in Madhya Pradesh. They have also laid down specific parameters for selecting the plant species. They have suggested different trees for specific localities in the factory taking into account the type of affluents emitted by the particular plant of the factory. Plantation raised by Bhor Industries, Vadodara, is another example pleading the role of plants in pollution abatement. Profuse plantation of *Casuarina equisetifolia, Eucalyptus* hybrid and *Leucaena leucocephala* was planned from 1981 to 1983 over an area of 2 hectares on the vacant strip between the factory and adjoining highway. Ahmad *et al.* (1988) have prepared a list of pollution tolerant trees shrubs and herbs growing indigenously round thermal power plants at Obra, Shakti Nagar and Renusagar in Uttar Pradesh.

ACKNOWLEDGEMENT

The author are thankful to Dr. (Mrs.) Shashi Rai, Additional Director, School Education, Bhopal for encouragement and to Dr. Ram Parasad, Chief Conservator of Forest, Bhopal for kindly going through the manuscript.

REFERENCES

1. Agrawal, A.L. 1988. "Air Pollution control studies and impact assessment of stack and fugitive emissions from CCI Akaltara Cement Factory". Project Report; Project sponsored by M/s. CCI Akaltara Cement Factory. Pub. by NEERI Nagpur.

2. Agrawal, A.L. 1989. "Air Pollution Control and impact on air environment for Mandhar Cement Factory". Project Report; Project sponsored by Cement Corporation of India, Mandhar Cement Factory Pub. by NEERI, Nagpur.

3. Ahmad, K.J. Yunus, M., Singh, S.N., Shrivastava, K., Singh, N. and Kulshrestha, K. 1988. Survey of Indian plants in relation to atmospheric pollution : A research project. In Perspectives in Environ. Bot. 2 : 283-306. Today and Tomorrow's Printers and Publishers, New Delhi.

4. Bennett, J.H. and Hill, A.C. 1973. Inhibition of apparent photosynthesis by air pollutants. *J. Environ. Qual.* 2 : 526-530.

5. Bernatsky, A. 1969. In Air pollution Proc. First European Congress on "The influence of air pollution on plants and animals". Wageningen, pp. 382-395.

6. Dubey, P., and Amritphale, D. (Eds.) 1984. Procedure Manual Water Pollution, Air Pollution and Mycotoxins Analysis. State Level workshop. Environ. Problems and Method to study them. Ujjain, M.P. (India).

7. Duxbury, A.C. and Yentsch, C.S. 1956. Plankton pigment monographs. *J. Marine Res.* 15 : 19-101.

8. Flaming, G. 1967. "When can forest belts reduce the emissions concentration". Luft and Kacltetchni K6, 255-258.

9. MacLachlan, S. and Zalik, S. 1963. Plastic structure chlorophyll concentration and free amino-acid composition of a Chlorophyll mutant of barley. *Can. J. Bot.* 41 : 1053-1062.

10. Rao, D.N., Agrawal, M. and Nandi, P.K. 1985. Urban Industrial Air Pollution and Plant life. In Perspective in Environmental Botany. Vol. I, Pub. Print House, Lucknow (India), pp. 189-210.

11. Singh, S.K. and Rao, D.N. 1983. Evaluation of plants for their tolerance to air pollution. Proc. Symp. on Air Pollution Control. Nov. 83, 218-224.

12. Spedding, D.J. 1969. *Nature* (London) pp. 224-229.

13. Tingey, D.T. 1968. M.A. Thesis, Botany Deptt. University of Utah, Salt Lake City, Utah.

14. Tiwari, S.L. 1991. Studies of Air Pollution Tolerance Indices of some planted trees in urban areas of Bhopal with reference to Eco-planning of industrial areas. Ph.D. Thesis, Barkatullah University, Bhopal.

15. Warren, L. 1973. Green space for air pollution control. N.C. State University Sch. for Resource. Tech. Rep. 50, Raleigh, N.C.

16. Weatherly, P.E. 1950. Studies in the water relations of the cotton plant. I. the field measurement of water deficits in leaves. New Phytol. 49 : 81-97.

ENVIRONMENTAL IMPACT ON THE FLORA OF GWALIOR AND CHAMBAL DIVISIONS OF MADHYA PRADESH

J.P. Kaushik

ABSTRACT

This chapter deals with effect of environmental conditions on the flora and floristic elements of Gwalior and Chambal divisions situated in Northern Part of Madhya Pradesh. The area is divisible in to South cooler and wetter and North warm and drier parts. The former is having higher percentage of Indian and eastern floristic elements while North part is dominated by elements of tropical and warmer elements. In general flora of the region is having composite types of floristic elements. The anthropogenic and developmental activities are changing the environmental conditions which are helping the exotic species to invade the area. How these "Neophytes" are changing the flora of the area has been discussed.

INTRODUCTION

There is a synchronised relationship between environment and organism embodied in it. Organisms are either fitted in or are filtered out by environmental parameters. But due to over exploitation of resources, pollution, etc. in modern technological society this dynamic system has become disbalanced, due to which large number of recorded wild flora and fauna species have extincted or are becoming as threatened species. This process is helping to accommodate new exotic species and in turn these are mounting a competitive stress on existing plant species at regional or local levels, due to which present species are being replaced by new ones either in terms of ecology or evolution. This phenomena is clearly visible in the area of North Madhya Pradesh. This part of the state is having great varieties of habitats, edaphic and climatic conditions and harbouring a variety of plant species. This area is experiencing a drastic qualitative and quantitative changes in its biodiversity (Kaushik, 1991 and 1994). Some 'ultra neophytes' such as *Parthenium hysterophorus* is monopolizing the area and reducing the density of other plant species (Kaushik, 1994). In this present communication an attempt is being made to describe the changing patterns in floristic elements composition and changes in plant species.

STUDY AREA AND ITS CLIMATE

The present study area comprises six districts viz., Guna, Shivpuri, Gwalior, Morena, Datia and Bhind of Madhya Pradesh covering an area of 474474 km^2. In past this area had luxurious semi-evergreen type of vegetation (Luard, 1912, Jain, 1978 and Kaushik, 1983). At present this area is having dry deciduous miscellaneous type of the vegetation of which Anogerssus pendula community is dominant and other 16 plant communities are derived from it (Kaushik, 1973).

The climate of this area is semi-arid monsoonic type experiencing long dry period. The districts of Guna and Shivpuri are cooler and having higher rainfall in comparison with remaining districts (Table 7.1).

Table 7.1 : Showing the average of various climatic characteristics of different districts of North Madhya Pradesh

District	*Temperature °C*		*Annual Rain-fall (mm)*	*Relative Humidity*	
	Mini.	*Max.*		*Mini.*	*Maxi.*
Morena	5.9-29.5	16.9-48.8	619.6	16.8	60.0
Gwalior	2.0-28.0	26.0-48.0	1060.0	20.0	85.6
Shivpuri	2.0-26.0	22.0-45.0	1100.0	24.0	89.0
Guna	2.0-23.0	23.0-43.0	1262.0	25.0	97.6

OBSERVATIONS

Characteristics of the flora

The Angiospermic species recorded from this area are 1209 belonging to 667 genera and 133 families. Out of these monocots are 270 species and remaining 939 species are dicots. The largest family of the area are Leguminous with 169 species followed by Graminae. The largest genus is *Cyperus* (14 species) followed by *Euphorbia* and *Ipomoea* (16 species each), *Eragrostics* (13 species), *Fimbristylis* (12 species), *Crotalaria* and *Cassia* (11 species), *Indigofera* and *Bluma* (10 species each) while 489 genera are represented by single species in this area. The flora of the area is herbaceous type. Near about 75 per cent species are herbaceous, 136 species are climber while 70 species are aquatic.

Floristic elements

The flora is having 57.15% of cosmopolitan floristic elements and 16.62% Indian elements but no species is endemic to this area. The percentage composition of floristic elements in the flora of different districts is variable (Table 7.2)

Table 7.2 : Showing 6 per cent distribution of different floristic elements in different districts of the area

	Indian	*Eastern*	*Warm countries*	*Cosmopolitan*
Morena	13.55	2.89	32.15	51.41
Shivpuri	13.80	6.6	22.2	57.4
Guna	15.15	16.15	13.51	55.15

Diminishing species from the area

It is interesting to note that due to anthropogenic activities some plant have become 'extinct' from the area which were occurring about 10-15 years back (Kaushik, 1983). These species are *Butea monosperma* var. *lutea*, *Caesalpia sappan*, *Striga gesnerioides* and *Ceratopteris thallictroides* (a Pteridophyte), besides these some species are also showing their diminishing trend due to (i) loss of habitats e.g. *Utricularia aurea* from Dhobital of Shivpuri (ii) over exploitation. Kaushik (1991) reported that *Gymna sylvestris, Zyziphus glaberrima, Boswellia serrata* are showing decreasing trends due to their over exploitation for various purposes. (iii) Stresses from *'ultra neophytes'*.

The area is occupied with some exotic plant species (Kaushik, 1973). The most common are *Acanthospermum huspidum, Agremone maxicana, Alternauthera sessilis, A. pungens, Cleome viscosa, C. gynandra, Cassia obtusifolia, Tephrosia hamilatonii, Lantana camara, Ipomoea fistulosa, Eichhornia crassipes*, etc. but human activities have changed the environmental conditions and created large number of micro-habitats which could accommodate some '*ultra neophytes*' (The term is used for those exotic species which have arrived here with last fifty years of less (Kaushik, 1994) such as *parthenium hysterophorus*, which is an aggressive weed increasing its number of plants per unit area and its distribution (Figs. 7.1-7.3)

DISCUSSIONS

The area in past was having semi-evergreen vegetation as is supported by paleological and historical studies of this area (Luard, 1914; Jain, 1978 and Kaushik, 1983) but due to anthropogenic and developmental activities are depleting the original vegetation and plant species, and are providing favourable conditions for the herbaceous flora as well as spread of exotic species (Jain, 1967). The climate of the area is showing seasonal variations are favourable for the composite type of floristic elements in the flora of this area (Maheswari, 1963 and Sharma *et al.*, 1988). The occurrence of elements of tropical and warmer countries in the flora of this area are due to higher temperature and prolonged dry conditions (Mehr-Homji and Misra, 1973).

The variation in present distribution of floristic elements in different parts of the study area require some explanation. In past this area had been occupied by temperate, Mediterranean European and Indo-Malayan elements because there was uniform vegetation of semi-evergreen type. But in due course of time due to climatic changes and human activities the area became delimited in warm drier and water and cooler areas (Table 7.1). It is due to this reason wetter and cooler part of the area (districts of Shivpuri and Guna) is having higher percentage of Indian and Eastern elements while warmer and drier part is having higher percentage of tropical and warmer elements (Agharkar and Ghosh, 1931).

The human interference has created large number of micro-habitats with less biotic competition. This changed situation is suitable to provide space for exotic species (Pichard, 1960) and the increased anthropogenic interference has further changed the situation in this area as since *Parthenium hysterophorus* has invaded this area is increasing its numerical strength and distribution in the different habitats of the area, marching towards monopolizing these habitats (Fig. 7.1-7.3), may be due to its higher reproductive capacity effective utilization of resources and better seed dispersal (Saxena and Ramakrishanan, 1982). The perusal of Tables 7.1-7.3 clearly indicates that those species which are perennial and establishing

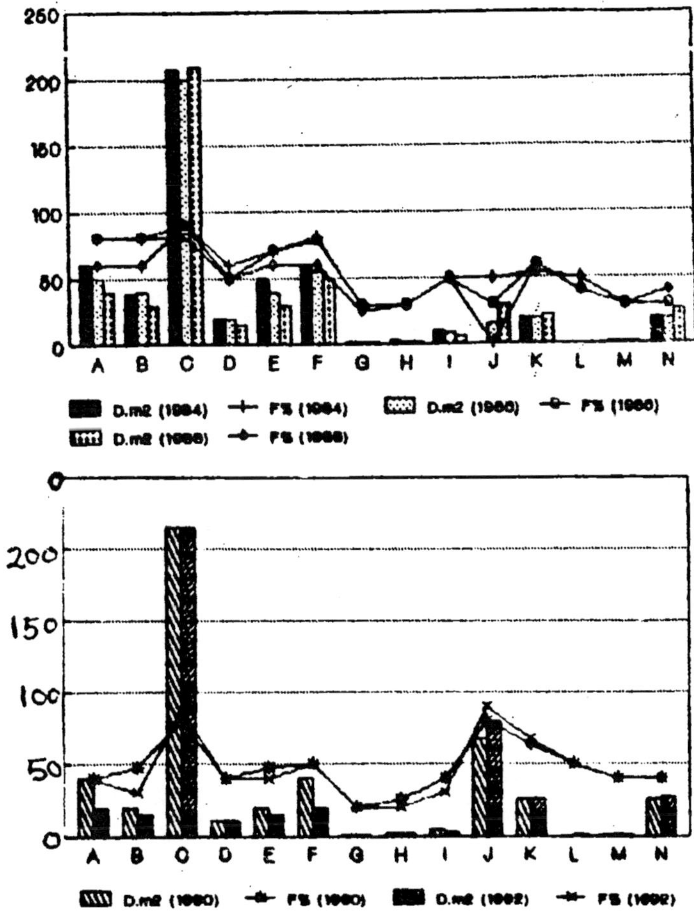

Fig. 7.1 Number and Frequency of Some Species on Fallow Lands of Gwalior

A.	*Cleome gynandra*	B.	*Cassia obtusifolia*
C.	*Bothriochloa pertusa*	D.	*Alysicarpus monilifer*
E.	*Crotolaria medicaginea*	F.	*Tephrosia hamiltonii*
G.	*Xanthium strumarium*	H.	*Croton bonplandianum*
I.	*Amaranthus spinosis*	J.	*Parthenium hysterophorus*
K.	*Convolvulus Pleuricaulis*	L.	*Blepharis maderaspatensis*
M.	*Boerhavia diffusa*	N.	*Merremia emerginata*

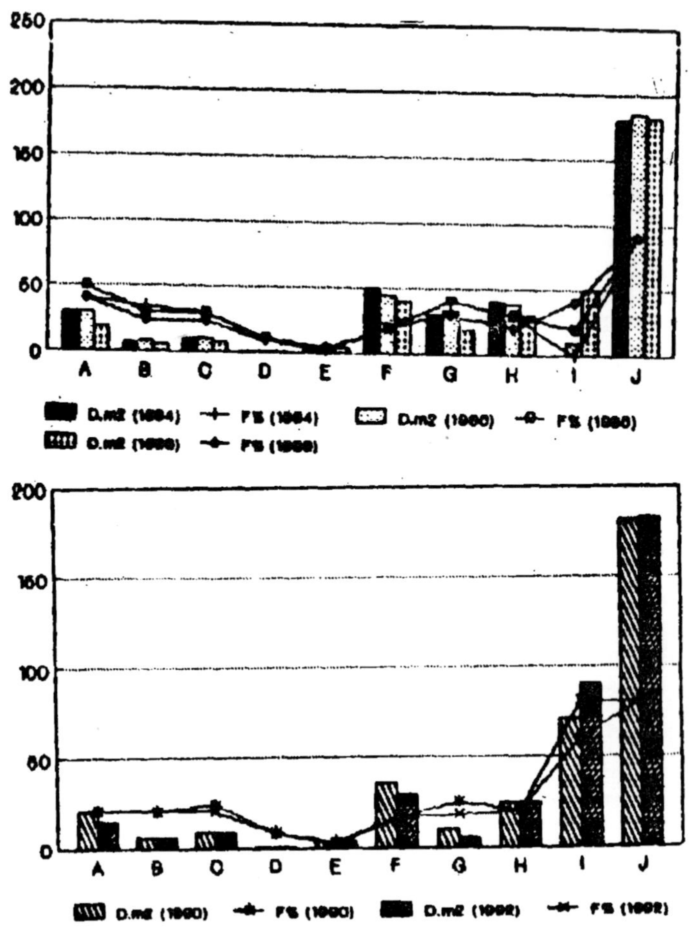

Fig. 7.2 Number and Frequency of Some Species on Ephemeral Water Bodies

A.	*Cassia obtusifolia*	B.	*Crotolaria medicaginea*
C.	*Tephrosia hamiltonii*	D.	*Euphorbia hirta*
E.	*Eclepta alba*	F.	*Echinochloa colonum*
G.	*Argemone maxicana*	H.	*Alternanthera sessiles*
I.	*Parthenium hysterophorus*	J.	*Cynodon dactylm*

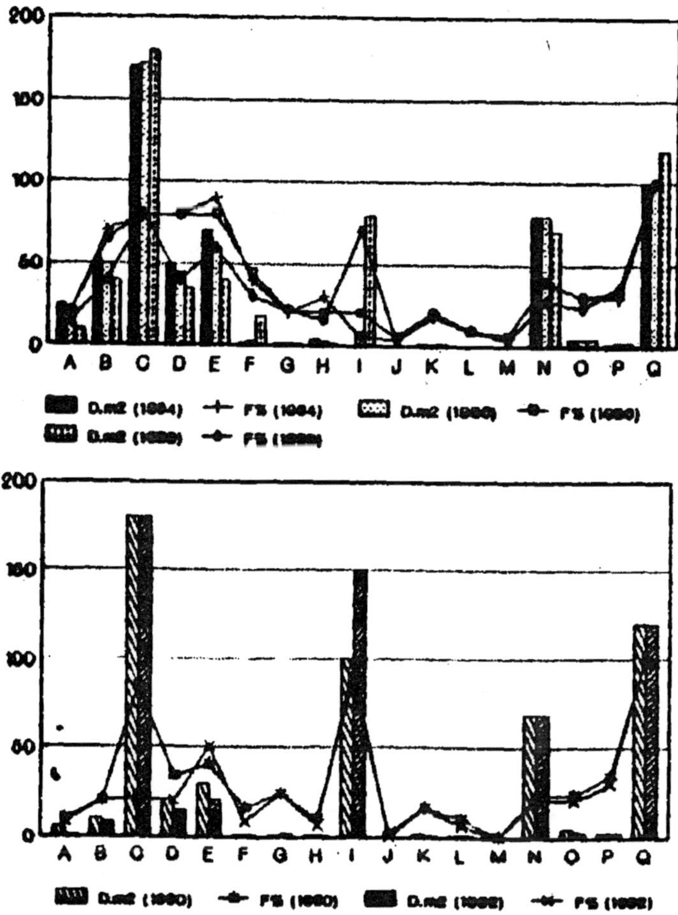

Fig. 7.3. Number and Frequency of Some Species on Disturbed Sites of Gwalior

A.	*Cleome gynandra*	B.	*Cassia obtusifolia*
C.	*Bothriochloa pertusa*	D.	*Crotolaria medicaginea*
E.	*Tephrosia hamilatonii*	F.	*Xanthium strumarium*
G.	*Croton bonploandianum*	H.	*Amaranthus spinosus*
I.	*Parthenium hysterophorus*	J.	*Tephrosia strigosa*
K.	*Tridex procumbanse*	L.	*Indigofera cordifolia*
M.	*Indigofera linifolia*	N.	*Aristida adscensionis*
O.	*Alternanthera pungens*	P.	*Heteropogon centortus*
Q.	*Cynodon dactylon*		

themselves through vegetative propagation in addition with seeds are increasing their population but annual species, erect in habit and establishing through seeds only are exhibiting a decreasing trend to reveal the exact mechanism of decreasing or increasing trends of different species and its impact on ecomorphological or genetical changes needs a intensive study.

ACKNOWLEDGEMENT

The author is highly thankful to Prof. R.R. Das, Head School of Studies in Botany for various suggestions and criticisms during the course of study and preparation of this paper.

REFERENCES

1. Agharkar, S.P. and A.K, Ghosh 1931. The composition of Bengal Flora (Abst.) *Proc. 18th Indian Sc. Cong*. Pt. II, 278.
2. Jain. A.K. 1978. *Study of the vegetation of certain area of Chambal ravines*. Ph. D. Thesis, Jiwaji University. Gwalior.
3. Jain, S.K. 1967. Phytogeographical considerations on the flora of Mount Abu. *Bull. Bot. Surv. India,* 9(1-4) : 68-78.
4. Kaushik, J.P. 1973. *Studies on the vegetation of Shivpuri and Karea Tehsils of District Shivpuri Madhya Pradesh*. Ph. D. Thesis, Jiwaji University, Gwalior.
5. Kaushik, J.P. 1983. *Flora of Shivpur,* M.P. Gwalior.
6. Kaushik, J.P. 1991. Flora of Central India : Some interesting aspects. *Botanical Researches in India,* 44-49 (Aery, N.C. and B.L. Chaudhary, Editors). Himansu Publ., Udaipur.
7. Kaushik, J.P. 1994. Effect of Environment on the flora of Gwalior region of North Madhya Pradesh (Accepted for Publication). *Act. Ecologica.*
8. Luard, C.I. 1912. *Gazetteer of Gwalior State,* Allahabad.
9. Maheswari, J.K. 1963. *Flora of Delhi*. CSIR New Delhi.
10. Meher-Homji, V.M. and K.C. Misra 1973. Phytogeography of Indian continent. *Progress* in *Ecology*. Vol. I (Misra, R. and J.S. Singh, Editors).
11. Pichard, T. 1960. Race formation in weedy species with special reference to *Euphorbia cyparissias* L. and *Hypericum* perforatum L. pp. 61-66 in J.L. Harper (Ed.). The *Biology* of *Weeds*. Black Well Scientific publ., Oxford.

12. Saxena, K.G. and P.S. Ramakrishnan 1982. Partitioning of Biomass and nutrients in secondary successional herbaceous population subsequent to slash and burns. *Proceedings of Indian Science Academy*, B.48, 807-818.

13. Sharma, M.,D.S. Dhaliwal, R. Gupta and P. Sharma 1988. Flora of Semi-arid Punjab India, Patiala.

CONSERVE ECOSYSTEMS IN TOTO*

H.K. Goswami

ABSTRACT

Based on surveys in the forests of Southern, Northern, Eastern and Central India during past three decades. The author suggests that all such forests which have survived, must immediately be totally conserved. A large number of endangered species of plants and animals can be regenerated and their germplasm conserved in these forests after due biological considerations.

Big Lakes in this country must also be protected as precious monuments in archaeology are protected. Wild relatives of plants, particularly medicinal plants suitable to each area should be planted around the lake in two or three tiers. Bhojwet land lake conservation plan in Bhopal is a rare multidisciplinary project setting such a high standard of Lake management and conservation.

Creation of "forest pockets" is also advocated which also refers to the maintenance of grass-lands for future crops. By and large, this entire planning must involve botanists and field biologists with positive and practical motives. We will have to conserve germplasm of plants and animals for our descendants.

INTRODUCTION

Our earth is a beautiful planet inhabited by hundreds of thousands of plants, animals including the most predominant, so called highly evolved "Homosapiens". This biped creature, linguistically designated as MAN is understood to have evolved from 0.5 to 0.2 million years ago. Last

---

* Based on Lectures delivered at the Institute of Nature Conservation, Krakow Poland and Institute of Zoology, University of Zurich, Switzerland during May - June, 1994.

few thousand years are very well depicted in archaeology, history and cultural documents of the progress of man. (Brodrick, 1971). Biologically this species *Homosapiens* to which we all human beings belong, has progressed immediately in attempting for control of biological functions and utilise each and every thing for own use/welfare. Excessive population growth and non-biological competitions have, however, led to total devastating results and we, at the end of 20th Century of modern Christian Era, are terribly burdened by innumerable problems. This is particularly more painful to realize that on the fronts of life-saving devices, we have gradually achieved remarkable success and nevertheless we countermand progress as a whole by "killing" devices. We want to eradicate others to make our life happy. Appearing as a philosophic statement, this is the dogmatic presentation of entire progress that we have made, "eradication of others" as a principal tool for our own survival and progress.

Biologically, survival is a biological priority of and for, each individual. But "progress-instincts" have no end. Hereunder, I have tried to analyse on biological grounds, why are we facing present environmental crisis and what actually we must plan for future before that it too late!

The Approach

1. We must know following things which have been proved to by dogmatically true though our modern scientific experimentations, these are:

(i) In Indian folk song in *Bundeli* (Dialect of Hindi), there is a saying "MAN NEEDS WOOD AT THE TIME OF BIRTH, GROWTH, NOURISHMENT, FOOD, CLOTHES AND AT DEATH". This is the bitter truth of life for 95 per cent human being on this earth, 5 per cent human who depend on "plastics"/"synthetic media" not on plant products, however cannot deny that "food" has to be dependent on plants. Some, even argue that animal food will be or can be sufficient, why grow so huge crop fields?

2. Every thing depend on some or the other thing. No "life" can lead life exclusively singly. Yes, this is another biological dogma with no exception. I had a querry from a young man in Washington (in 1991), "why not store capsules with vitamins, proteins and food items and preserve for a country? Modern science can do that, why cultivate such huge paddy/maize/wheat fields"? I replied, yes, this is possible for one or say a few thousand families but what about millions and millions? Food problem is

a World Problem. For example, what is happening in South Africa: Millions are annually starved to diseases and death! Furthermore, suppose you develop a disease at the age of 60 or some one is born in your family who is genetically incompetent to utilise sugar, carbohydrates, proteins or other things, what should be his/her fate? Certainly, we have to arrange for special food. Pests, moths, thousands of species of poisonous insects and other animals are covered, accommodated, tolerated and, or, their presence is balanced by forests, crop fields and other cultivated plants! What will happen if there be no forests? You will eat capsules and live with an oxygen cylinder, masked within a plastic cage?

So, precisely no life form, including man can live without animals and more elaborately and emphatically, without plants.

Yet another argument will be a final blow for such a thought. What about natural disastrous calamities? Do you know, lands and mountains have shifted, migrated, moved, sunk and teared apart in the past and will do so in future too? If the land used for your multi-trillion dollar plastic cased capsules and food supplies sunk by a single stroke of earth-quake or sea-storm, what would the surviving population do? Actually, nature offers substitute to nature, man can only manipulate to some extent.

Why conserve biodiversity? Why not offer priority and kill others?

This is a very fashionable objection by pseudo-modern thinkers and planners. During discussions in various International Congresses on Environmental Protection etc., people pose this as a possible solution for more use of land for purpose other than growing plants/forests. I straightway ask them - who will chose what species you should kill and what should be preserved? Your list may be suitably prepared for today; how do you know that it would be valid for years to come? Everything changes or is prone to change, the DNA (genes), the chromosomes, the cells and tissues, organs and receptive conditions, their adaptability, mode of functioning, everything is prone for a change! There are hundreds of diseases, yet unknown and there are thousands of plants and animals totally unexplored. Modern radar, air-crafts, fighters, space shuttles are based on combination and utilisation of physical science with variety of plant and animal lives. So people can always spend millions of dollars on preparing a list of plants and animals to be "eradicated" in order to "vacate" the land to be used for modern purposes, such attempts will be futile and always incomplete, however! Interdependence of plant and animal species are not yet fully documented.

Biodiversity has evolved in nature to counter-balance the survival of other species, related and un-related and in majority, each of them is an integral part of a biological net-work.

Some Examples of Rare Significance

There are a few major problems of great significance on maintenance of biodiversity.

(A) *What do we get?* Almost everything! The total system constitutes usefulness for human being. The presence of one species is dependent on the other; for pollination of trees and fruit dispersal we need besides air, birds insects, monkeys; for good soil we need bactera and various micro-organisms for animals, their reproduction, survival, we need ground flora, water and bushes. Entire *Ayurvedic* medicines are derivatives of essentially plant products; most drinks, fruits and vegetables can never be substituted by synthetic products.

(B) *What do we loose, if we do not conserve ecosystem?* The number of species that have been declining every year and which have been identified to exhibit degradation of our ecosystems are solely due to our deliberate negligence. Over 3000 plants species are endangered in our country. The world's top biologists warn that unless something is done by the Governments to reserve the process, the world will lose many species that almost certainly could have provided new drugs, new good sources and many other opportunities for economic development.

This is most surprising that 50% of medicinal plants are still found in wild and exclusively form core for the modern medicine in Asia, Africa and Latin America. So if we continue to loose forests, grasslands and aquatic ecosystems we ought to bankrupt our biological economy for future.

(C) *New discoveries in Nature:* Besides regular exploitation by us for our needs, we must and will have to depend for future on total ecosystems. There are a large number of plants and animals which have arisen as a genetic consequence in nature. We have recently heard of a discovery of a new bird by Paul Salman from the University of East Anglia in the rain forest of the Colombian Andes in 1991. This bird, not yet named, is golden in colour and looks too pretty and feeds on insects. Our own work in Madhya Pradesh has resulted in discovering a few wild species of *Isoetes, Ophioglossum* and *Hibiscus* which are of immense medicinal significance. There may be hundreds of this kind in India which could be detected on account of an "intact" ecosystem, the best example is a series of such discoveries from "Silent Valley" in Kerala. Hundreds and thousands are yet to be discovered from Indian sub-continent.

Plan for Germplasm Conservation

We must now be very careful in shaping and utilising the natural resources with a very well-biological view point. We need large grass-fields, wild forests, protected and almost "sealed" so that future evolution of cereals, legumious plants, insects, rodents and lower mammals continues to take their own course. By artificial hybridisation, we can produce maximum exploitation of genes but this potential exploitation will cease after a few hundred years if we do not have raw genetic material in the gene pool.

Therefore, world over, there is a serious thought that we must conserve the wild relatives of all possible plants and animals. The best way is, which is also the first step to conserve, protect all surviving forests *in toto* without fail and without any "ifs and buts". In these protected forests, we must plan to conserve some of the endangered species of plants and animals. We all know, this is possible.

All existing lakes in the country should be taken care as much as the Archaeology department caters to the architectural and historical needs of a monument. Wild relatives of the pertinent flora must be planted around the fringe area of each lake. Our conservation plant executed in Bhopal (Bhoj wet land) will go a long way in the history of lake conservation. In my view, which is based on extensive travelling from South to North and from East to West in this country by road and rail during 30 years, we must plan to develop forest pockets locally suitable in the barren area. This first preference for biological conservation in an area must be given to the local flora and fauna. Second choice should be again based on biological need and not on the whim of Planners or Politicians. We must immediately stop building big dams by cutting long belt of forests. The benefits of today based on destruction of ecosystems will sow the seeds for tomorrow's devastations. Huge floods, earthquakes, deserts and diseases epidemics are some of the best examples based on breakdown of ecosystems.

REFERENCES

1. Bir, S.S., 1987. Pteridophytic flora of India: Rare and Endangered Elements and their Conservation. *Indian Fern J.* 4 : 95-101, 1987.

2. *Red Data Book on Polish Plants*. 1993. Published by Polish Academy of Sciences, Warsawa/Krakow.

3. *Red Data Book on Indian Plants*.

4. Goswami, H.K. 1987. Biological impact of Industrial Hazard: A reference to Bhopal Gas Tragedy, In, *comparative Environmental Mutagenesis*, Ludhiana.

5. Goswami, H.K. 1992. Must we learn lessons from Bhopal Gas Tragedy. In, *Industrial Hazards*, New Delhi.

6. Goswami, H.K. 1992, 1993. Reports on Bhojwetland, E.P.C.O., Bhopal (A monographic treatise on Biology of upper Lake *Bionature* 14 (2), 1994.

INHIBITION OF NET PHOTOSYNTHESIS IN SOYBEAN BY SO_2 AND NO_2 APPLIED ALONE AND IN COMBINATION

S.N. Singh

ABSTRACT

Phytotoxicity of individual air pollutants SO_2 and NO_2 is well recognised. However, information available on synergism of these pollutants is sporadic and inadequate. To investigate their synergistic action on plant growth, soybean plants were grown in field conditions and intermittently fumigated by 1.0 ppm SO_2, 1.0 ppm NO_2 and 0.5 ppm SO_2 + 0.5 ppm NO_2 separately, daily for 1 h. The plant samples were periodically analysed for foliar injury, chlorophyll concentration, carbohydrate level and phytomass accumulation. The results showed that SO_2, when applied alone, caused to appear scattered chlorotic and necrotic lesions, destroyed chlorophyll pigments and decreased the carbohydrate level and phytomass accumulation. Individually, NO_2 also showed the similar effects, barring an initial increase in photosynthetic pigments. In NO_2-exposed plants, streaks developed along the mid-rib of the leaves, but with combined treatment of SO_2 + NO_2, the plants displayed mixed foliar injury symptoms and reflected more reduction in chlorophyll and carbohydrate contents as well as in dry matter accumulation than that exposed to SO_2 or NO_2 alone. This indicates that interaction of SO_2 and NO_2 is more phytotoxic than the action of individual pollutants.

INTRODUCTION

The plants growing in natural conditions are concomitantly exposed to an array of air pollutants, usually emanating from fossil fuel combustion. These pollutants interact together and affect the plant metabolism in a way different; the effect of pollutant mixture may be either synergistic, additive or antagonistic, depending upon various plant and environmental factors (Tingey and Reinert, 1975).

Although, the toxicity to plants caused by their exposure to single pollutant in the laboratory conditions had been already studies in details, the work on the effect of pollutant mixtures was initiated as early as in 1970. Dochinger (1970) showed that an exposure to a mixture of SO_2 and O_3 produced injury symptoms in eastern white pine at a concentration much lower than the threshold for SO_2 or O_3 alone. A reduction in CO_2 uptake by *Picea abies* exposed to SO_2 and HF simultaneously was reported by Keller (1980). On the contrary, Kress (1980), who exposed ten tree species to $NO_2 + O_3$, found less injury in them than that when exposed to SO_2 and O_3 separately. Matsushima (1971) observed more severe injury in moong and tomato plants fumigated with PAN and SO_2 together than with PAN or SO_2 alone. An evidence of synergism between SO_2 and NO_2 in their combined effects on plants and in lowering of the threshold limits required to produce injury by either pollutant, was produced by the studies of Tingey *et. al.* (1971) and Skelly *et. al.* (1972). Hill and Bennett (1974) reported an inhibition in apparent photosynthesis of alfalfa plants when exposed to a combination of SO_2 and NO_2. Ashenden (1979) noted a short term increase in transpiration rate of two grass species when SO_2 and NO_2 were applied alone, but a decrease when both the pollutants were applied together. Amundson and Weinstein (1980) observed that an exposure of soybean plants to SO_2 and NO_2 together induced early senescence and suppressed yield in comparison to those exposed to SO_2 or NO_2 alone.

Since SO_2 and NO_2 both are commonly present in our urban and industrial atmospheres, the present investigation was undertaken to study the effects of these pollutants alone and also in combinations on chlorophyll pigments, carbohydrate level and phytomass accumulation of *Glycine max*.

MATERIALS AND METHODS

Viable seeds of *Glycine max* Merr (cv. Clark-63) were soaked overnight in deionized water and sown in four one meter square plots (25 seeds in each plot) using farm manure (5 kg per plot). The seeds germinated at 25 °C in daylight with a relative humidity of 60-75%. The plants of plots 1, 2 and 3 were fumigated with 1.0 ppm SO_2, 1.0 ppm NO_2 and 0.5 ppm SO_2+0.5 ppm NO_2, respectively, for 1 h every day in the morning hours, inside the polythene chambers supported on $1m^3$ iron frames as scheduled in Table 9.1. The treatment was given to plants between 30-d and 80-d ages. During fumigation, mini-fans were provided inside the polythene chambers to prevent the formation of boundary layer resistance and to facilitate proper mixing of gases in the chambers. However, the plants of plot 4 remained untreated to serve as control for all sets of fumigation.

Table 9.1 : Fumigation and sampling schedule of *Glycine Max*

		Cumulative dose (c × t)			
Plant age (days)	*Experimental condition*	SO_2 (1.0 ppm)	NO_2 (1.00 ppm)	$SO_2 + NO_2$ (0.5 ppm + 0.5 ppm)	*Sampling order*
1-30	Normal	-	-	-	-
40	Fumigation	10	10	10	I
50	-do-	20	20	20	II
60	-do-	30	30	30	III
70	-do-	40	40	40	IV
80	-do-	50	50	50	V
80-90	No fumigation	50	50	50	VI

Generation of pollutants

SO_2 was generated chemically by the reaction of dilute H_2SO_4 with known amount of 1 per cent aqueous solution of sodium sulphite in a flask (Rao and LeBlanc, 1966) to give 1.0 ppm SO_2 concentration in 1 m^3 volume of air inside the polythene chamber. Similarly, NO_2 was produced by the reaction of known amount of 1% aqueous solution of sodium nitrite and dilute HNO_3, to obtain 1.0 ppm NO_2 concentration in the same fumigation chamber. For combined treatment of plants with SO_2 and NO_2, half the amounts of sodium sulphite and sodium nitrite were reacted with dil. H_2SO_4 and dil. HNO_3, respectively, in different flasks to produce 0.5 ppm concentration of each gas in the fumigation chambers.

Collection of plant samples

The plant samples were collected periodically 40, 50, 60, 70, 80 and 90-d ages from each plot and analysed for foliar symptoms, chlorophyll concentration, carbohydrate content and phytomass accumulation.

Chlorophyll

Chlorophyll pigments were extracted from leaves in 80% (v/v) acetone and centrifuged at 3000 g for 15 min to remove the debris. The volume of the clear extract was made up to 100 ml by the addition of 80 per cent acetone and its observance at 643 and 663nm was measured with a spectrophotometer Model SP-500 UNICAM. The concentration of chlorophyll was determined following the formulae given by Maclachlan and Zalik (1963).

Carbohydrate

The carbohydrate level was determined in the oven-dried and powered plant samples following the enthrone method (Plummer, 1971).

Phytomass

The intact plants were oven-dried at 80°C for 24 h and then weighed till constant weight was obtained to set phytomass values in g/plant.

RESULTS AND DISCUSSION

Foliar surface of *Glycine max* displayed characteristics injury symptoms, when exposed to SO_2 and NO_2 alone and in combination for 15 days; SO_2- exposure developed sporadic spots, while NO_2- fumigation produced streaks along the mid-rib of the leaves. In the case of $SO_2 + NO_2$ treatment, symptoms borne by the leaves were amalgam of two types as produced by SO_2 and NO_2 individually. With increase in pollutant-dose, light green chlorotic regions turned brownish and necrotic, indicating the complete death of leaf tissues. It was also observed that the leaves of NO_2-treated plants became more green than that of control in the initial stage of fumigation.

Foliar injury symptoms are manifestations of phytotoxic behaviour of SO_2 and NO_2 (Hill et al., 1974; Srivastava et al., 1975). The appearance of localized symptoms in SO_2-treated plants clearly indicates that SO_2, on being readily dissolved in H_2O, is not transported through transpirational pull (MacLean *et al.*, 1968) and so causes local injury by the formation of SO_3^{2-} and SO_4^{2-} ions (Singh *et al.*, 1988). It is the fact that the toxicity of SO_2 is determined by the pH which controls the formation of SO_3^{2-} and SO_4^{2-} ions. According to Thomas *et al.* (1952), SO_3^{2-} ions are many times more toxic than SO_4^{2-} ions. Similarly, NO_2, on dissolution in water inside the leaf fissures, forms a stoichiometic mixture of NO_3^- and NO_2^- ions (Jolly,

1964). These ions unusually get detoxified and metabolised by the plants. But if not detoxified; they accumulate in the tissues to toxic levels and thereby causing injury to plants (Zeevaart, 1976). At the same time, it is presumed that initial metabolism of NO_2 by plants caused the NO_2-treated leaves to appear darker green than the control ones. In the case of SO_2 + NO_2-treated plants, both pollutants interacted together to produce combined injury symptoms on the leaves.

Table 9.2 : Changes in chlorophyll concentrations of *Glycine max* exposed to SO_2, NO_2 and SO_2 + NO_2 (average of 3 replicates)

Plant age (days)	*Control*	*SO_2 (1.0 ppm)*	*NO_2 (1.0 ppm)*	*SO_2 + NO_2 (0.5 ppm + 0.5 ppm)*
40	100	95.0	110.0	86.6
50	100	91.8	95.7	82.2
60	100	87.6	92.5	76.2
70	100	85.8	90.6	73.4
80	100	84.2	88.3	72.5
90	100	79.0	89.5	70.4

Values express as % control.

The content of chlorophyll was invariably lower in SO_2 and SO_2 + NO_2-fumigated plants than that in control (Table 9.2). However, the plants treated alone with NO_2 showed an initial increase in chlorophyll concentration which, perhaps, gave dark green appearance to the leaves. As the NO_2-dose increased with further fumigation, the chlorophyll amount decreased in the NO_2-treated plants also. This proves the phytotoxic nature of NO_2. At 40-d age, NO_2 exposed plants showed an increase of 10%, whereas SO_2 and SO_2 + NO_2-treated plants exhibited reductions of 15.0 and 13.4 per cent, respectively (Table 9.2). This indicates that the combined effect of SO_2 and NO_2 (0.5 ppm SO_2 + 0.5 ppm NO_2) is less than that of SO_2 (1.0 ppm) alone in early stage of fumigation. Despite of deleterious effects of the pollutants, the chlorophyll content continued to augment with plant age both in control and treated plants up to 70-d age and then declined. As compared to control, the magnitude of decrease in chlorophyll concentration amounted to be 21.0 in SO_2, 10.5 in NO_2 and 29.6% in SO_2 + NO_2-exposed plants of 70-d age at cumulative dose of 40 (c × t) of each pollutant. This observation adumbrates that the synergistic effect of SO_2 + NO_2 is more than that of SO_2 or NO_2 alone, but less than summed effects of individual pollutants.

Initial increase in chlorophyll amount with NO_2-fumigation has been also reported by Taylor and Eaton (1966). Probably, nitrate enrichment by NO_2-fumigation stimulated the synthesis of amino-acids, such as glycine which being skeleton of chlorophyll molecules, enhanced the chlorophyll concentration in NO_2-treated plants (Spierings, 1971; Yoneyama and Sasakawa, 1979). However, higher dose of NO_2 hampered the chlorophyll synthesis. An invariable decrease in chlorophyll content in SO_2-fumigated plants was due to chlorophyll degradation to pheophytin which no longer serves as photosynthetic pigments (Rao and LeBlanc, 1966; Coker, 1967). Interaction of SO_2 and NO_2 in the combined treatment of plants magnified their effect on chlorophyll pigments (Wellburn *et al.*, 1972).

Table 9.3 : Changes in total carbohydrate levels of *Glycine max* exposed to SO_2, NO_2 and SO_2 + NO_2 (average of 3 replicates)

Plant age (days)	*Control*	*SO_2 (1.0 ppm)*	*NO_2 (1.0 ppm)*	*SO_2 + NO_2 (0.5 ppm + 0.5 ppm)*
40	100	95.4	102.8	96.8
50	100	90.7	92.6	89.6
60	100	87.4	89.7	83.4
70	100	81.7	87.3	78.6
80	100	74.8	81.9	69.7
90	100	77.6	84.8	74.2

Values express as % control.

With respect to control, total carbohydrate level was lowered in all sets of treated plants, excepting an initial increase in NO_2-fumigated plants at 40-d age (Table 9.3). Perhaps, induced synthesis of chlorophyll pigments at lower doses of NO_2 accelerated the carbohydrate accumulation via photosynthesis. At higher doses of fumigation, the reductions in carbohydrate levels of SO_2, NO_2 and SO_2 + NO_2-treated plants were in the order of 25.2, 18.1 and 30.3 per cent, respectively, when pollutant dose culminated to 50 (c × t) in each case. This clearly reflects that the combination of SO_2 and NO_2 (0.5 ppm SO_2 + 0.5 ppm NO_2) is more phytotoxic than SO_2 (1.0 ppm) or NO_2 (1.0 ppm) alone. It may be inferred that the reduced rate of photosynthesis in treated plants due to leaf injury (chlorosis and necrosis), reduction in chlorophyll concentration and enhancement in respiration (Syratt and Wanstall, 1969) led to decreased accumulation of carbohydrate (Kondo and Sugahara, 1978).

Table 9.4 : Changes in phytomass accumulation of *Glycine max* exposed to SO_2, NO_2 and SO_2 + NO_2 (average of 5 replicates)

Plant age (days)	*Control*	*SO_2 (1.0 ppm)*	*NO_2 (1.0 ppm)*	*SO_2 + NO_2 (0.5 ppm + 0.5 ppm)*
40	100	95.9	105.2	96.4
50	100	80.7	92.1	87.6
60	100	81.7	83.8	74.2
70	100	72.5	79.6	67.9
80	100	67.6	74.8	63.8
90	100	69.8	77.8	65.8

Values express as % control.

The dry matter continued to accumulate in all sets of plants till 80-d age and then declined. However, the phytomass values of treated plants were invariably lower than that of control plants (Table 9.4). With respect to control, the phytomass values were reduced by 32.4, 25.2 and 36.2% in SO_2, NO_2 and SO_2 + NO_2-exposed plants, respectively, at 80-d age when the cumulative dose culminated to 50 (c × t) for each pollutant. Such a reduction may be attributed to several factors including leaf injury, chlorophyll degradation and other disturbances in metabolic processes (Sij and Swanson, 1974; Bull and Mansfield, 1974). Again it is apparent that the combined effect of SO_2 or NO_2 alone, but less than summed effects of individual pollutants, though the pollutant dose remained constant in all treated plants (Tingey *et al.*, 1971; Skelly *et al.*, 1972).

The above results evidenced to synergism between SO_2 and NO_2 which causes more damage of plants than SO_2 or NO_2 alone.

REFERENCES

1. Amundson, R.G. and Weinstein, L., 1980. The effects of SO_2 and NO_2 alone and in combination on yield of soybean. *Plant Physiol.,* 65 : 152.

2. Ashenden, T.W., 1979. The effects of long-term exposures to SO_2 and NO_2 pollution on the growth of *Dactylis glomerata* L. and *Poa pratensis* L. *Environ. Pollut.,* 18 : 258-259.

3. Bull, J.N.B. and Mansfield, T.A., 1974. Photosynthesis in leaves exposed to SO_2 and NO_2. *Nature,* Lond., 241 : 443-444.

4. Coker, P.D., 1967. Effects of sulphur dioxide on bark epiphytes. *Transactions of the British Bryological Society,* 5 : 341-347.

5. Dochinger, L.S., Bender, F.W., Fox, F.O. and Heck, W.W., 1970. Chlorotic dwarf of eastern white pine caused by ozone and sulphur dioxide interaction. *Nature,* 24 : 225-276.

6. Jolly, W.L., 1964. The inorganic chemistry of nitrogen. New York, W.A. Benjamin Inc.

7. Keller, T., 1980. The simultaneous effect of soil-borne NaF and air pollutant SO_2 and CO_2-uptake and pollutant accumulation. *Oecologia,* 44: 283-285.

8. Kondo, N. and Sugahara, K., 1978. Changes in transpiration rate of SO_2-resistant and sensitive plants with SO_2 fumigation and the participation of abscisic acid. *Plant & Cell Physoil.,* 19 : 365-373.

9. Kress, L.W., 1980. Effects of O_3 and $O_3 + NO_2$ on growth of tree seedlings. Proc. Symp. on "Effects of Air Pollutants on Mediterranean and Temperate Forest Ecosystems" 239. Rept. PSW-43, Pacific South-west Forest and Range Experiment Station, Berkeley, California.

10. Maclachlan, S. and Zalik, S., 1963. Plastic structure, chlorophyll concentration and free amino-acid composition of a chlorophyll mutant of barley. Can. J. Bot., 41 : 1053-1062.

11. MacLean, D.C., McCune, D.C., Weinstein, L.H., Mandl, R.H. and Woodruff, G.N., 1968. Effects of acute hydrogen fluoride and nitrogen dioxide exposures on citrus and ornamented plants of Central Florida. Environ. Sci. Tech., 2 : 444-449.

12. Matsushima, J., 1971. On composite harm to plants by sulphurous acid gas and oxidant. *Indust. Poll. & Damage,* 7 : 218-224.

13. Plummer, D.T., 1971. An introduction to practical chemistry. McGraw-Hill Book Co., England.

14. Rao, D.N. and LeBlanc, F., 1966. Effects of SO_2 on the lichen algae with special reference to chlorophyll. *Bryologist,* 69 : 69-75.

15. Sij, J.W. and Swanson, C.A., 1974. Short-term kinetic studies on the inhibition of photosynthesis by sulphur dioxide. *J. Environ. Qual.,* 3 : 103-107.

16. Singh, S.N., Yunus, M., Kulshreshtha, K., Srivastava, K. and Ahmad, K.J., 1988. Effect of SO_2 on growth and development of *Dahlia rosea* Cav. *Bull. Environ. Contam. Toxicol.,* 40 : 743-751.

17. Skelly, J.M., Moore, L.D. and Stone, L.L., 1972. Symptom expression of eastern white pine located near a source of oxides of nitrogen and sulphur dioxide. *Pl. Dis. Reptr.,* 56, January, 1972.

18. Spierings, F.H.F.G., 1971. Influence of fumigations with NO_2 on growth and yield of tomato plants. *Neth. J. Pl. Pathol.,* 77 : 194-200.

19. Srivastava, H.S., Jolliffe, P.A. and Runechles, V.C., 1975. The influence of nitrogen supply during growth on the inhibition of gas exchange and visible damage to leaves by NO_2. *Environ. Pollut.,* 9 : 35-47.

20. Syratt, W.J. and Wanstall, P.J., 1969. The effect of sulphur dioxide on epiphytic bryophytes. *Proc. Eur. Congr. Air Pollut.* Ist 1968, pp. 79-85.

21. Taylor, O.C. and Eaton, F.M., 1956. Suppression of plant growth by nitrogen dioxide. *Plant Physiol.,* 41 : 132-135.

22. Thomas, M.D., Hendricks, R.H. and Hill, G.R., 1952. Some injuries in the air and their effects on plants, pp. 41-47. In : McCabe, L.C. (Ed.), Air Pollution. *U.S. Tech. Conf. Proc.,* McGraw Hill, N.Y.

23. Tingey, D.T., Reinert, R.A., Dunning, J.A. and Heck, W.W., 1971. Vegetation injury from the interaction of nitrogen dioxide and sulphur dioxide. *Phytopathology,* 61 : 1506-1511.

24. Wellburn, A.R., Majernik, O. and Wellburn, F.A.M., 1972. Effects of SO_2 and NO_2 polluted air upon the ultrastructure of chloroplasts. *Environ. Pollut.,* 3 : 37-49.

25. White, K.L., Hill, A.C. and Bennett, J.H., 1974. Synergistic inhibition of apparent photosynthesis rate of alfalfa by combinations of sulphur dioxide and nitrogen dioxide. *Environ. Sci. & Tech.,* 8 : 574-576.

26. Yoneyama, T. and Sasakawa, H., 1979. Transformation of atmospheric NO_2 absorbed in spinach leaves. *Plant & Cell Physiol.,* 1 : 263-273.

27. Zeevaart, A.J., 1976. Some effects of fumigating plants for short periods with NO_2. *Environ. Pollut.,* 11 : 97-108.

EFFECT OF GRAPHITE FACTORY EMISSIONS ON THE NITROGEN AND PROTEIN CONTENTS OF SOME TREES

Dr. Smt. Swati Shrivastava
and M.P. Singh

ABSTRACT

In the present investigation the effect of Hindustan Electrographites Ltd., Mandideep (HEG) emissions on the nitrogen and protein contents of *Azadirachta indica, Bauhinia purpurea, Cassia fistula, Dalbergia sissoo, Eucalyptus citriodora, Ficus benghalensis, Ficus religiosa, Zizyphus jujuba* was studied. The amounts of total nitrogen and total Protein in all the above eight tree species growing in the effected area decreased due to HEG emissions, the amount of nitrogen and protein was less in the leaves of trees under invcstigation.

In affected trees of *Azadirachta indica, Cassia fistula, Dalbergia sissoo, Ficus benghalensis, Ficus religiosa* and *Zizyphus jujuba,* the amounts of total nitrogen and total protein were highest during the rainy season, the same decreased during winter season and increased in the summer season. In *Bauhinia purpurea* and *Eucalyptus citriodora* the amounts were highest during rainy season, then a decrease was noticed during the winter as well as the summer season. The lowest amounts were recorded in the summer season.

INTRODUCTION

The factory of Hindustan Electrographites·Ltd., is situated at Mandideep. Mandideep is located in South West of Raisen district on the national Highway no. 12 at a distance of about 25 kms. from the city centre of Bhopal. HEG is engaged in manufacturing graphite electrodes and other graphite specialities. During the manufacturing processes emitted pollutants are petro vapours heat, gaseous pollutants and graphite particles. The emissions drift with wind current and settle down on the surrounding vegetation. The area around the factory was taken as the polluted area.

A control area namely Ahmadpur nursery was selected 12 km. away from the factory. This area is not affected by factory emissions.

Nitrogen is one of the main structural components of the living organisms. As nitrogen plays a fundamental role in metabolism, growth, reproduction and heredity, it has a central place in the phenomena of life. Proteins are nitrogen containing molecules and are the building blocks of protoplasm.

In present investigation, the effect of HEG factory emissions on the total nitrogen and total protein contents of tree species were studied.

Eight plant species were selected for the study namely *Azadirachta indica, Bauhinia purpurea, Cassia fistula, Dalbergia sissoo, Eucalyptus citriodora, Ficus benghalensis Ficus religiosa* and *Zizyphus jujuba*. These plants are present in control as well as in polluted area.

The climatic conditions and the soil of the polluted area and that of the control area were nearly the same.

MATERIAL AND METHODS

Nitrogen contents were estimated from the oven dried leaf samples. The samples were powdered and nitrogen estimation was done by micro kjeldhal method (Jackson, 1962).

The total protein contents were estimated from the amount of total nitrogen present in the sample by the following formula-

100 parts of protein contained 16 parts of nitrogen.

The rate of became 100/16 = 6.26

i.e. in every 6.25 part protein, one part is nitrogen.

RESULTS

Average amount of total nitrogen (mg g^{-1}) in the polluted and control area is given in Table 10.1.

Average amount of total protein is presented in Table 10.2.

Table 10.1 : Average amount of nitrogen (mg g^{-1}) in the leaves of various trees of effected and control area

S. No.	Name of the species	Effected Control	AVERAGE AMOUNT OF NITROGEN (mg g^{-1})											
			Jul. 1988	Aug. 1988	Sep. 1988	Oct. 1988	Nov. 1988	Dec. 1988	Jan. 1989	Feb. 1989	Mar. 1989	Apr. 1989	May 1989	Jun. 1989
1.	*Azadirachta indica*	Effected	3.97 ±0.12	4.03 ±0.03	4.21 ±0.06	5.17 ±0.08	4.29 ±0.05	3.26 ±0.04	2.06 ±0.08	2.22 ±0.04	3.26 ±0.08	3.26 ±0.10	3.26 ±0.07	3.57 ±0.05
		Control	4.21 ±0.08	4.13 ±0.07	5.01 ±0.45	5.65 ±0.11	5.01 ±0.10	4.29 ±0.07	3.10 ±0.12	3.41 ±0.02	3.33 ±0.16	4.05 ±0.19	4.21 ±0.11	4.29 ±0.03
2.	*Bauhinia purpurea*	Effected	3.89 ±0.64	3.02 ±0.05	2.86 ±0.54	3.73 ±0.03	4.05 ±0.08	3.02 ±0.66	2.70 ±0.10	2.22 ±0.07	1.82 ±0.03	1.43 ±0.12	3.49 ±0.23	4.13 ±0.54
		Control	3.97 ±0.23	4.05 ±0.11	4.21 ±0.20	4.53 ±0.11	5.09 ±0.04	4.53 ±0.16	3.41 ±0.24	3.10 ±0.06	2.94 ±0.05	2.86 ±0.02	3.57 ±0.02	4.21 ±0.20
3.	*Cassia fistula*	Effected	3.02 ±0.03	3.89 ±0.06	3.57 ±0.19	4.37 ±0.10	4.05 ±0.02	2.38 ±0.11	1.43 ±0.23	2.78 ±0.16	3.49 ±0.10	3.32 ±0.11	3.81 ±0.40	3.10 ±0.23
		Control	3.65 ±0.04	3.97 ±0.11	4.29 ±0.06	4.93 ±0.06	4.77 ±0.07	3.41 ±0.20	2.94 ±0.19	3.26 ±0.04	3.65 ±0.03	3.31 ±0.07	3.97 ±0.20	3.89 ±0.12
4.	*Dalbergia sissoo*	Effected	3.26 ±0.15	4.05 ±0.11	4.37 ±0.08	2.14 ±0.21	2.22 ±0.07	2.06 ±0.14	1.66 ±0.11	2.54 ±0.03	3.02 ±0.04	3.49 ±0.11	3.57 ±0.06	2.70 ±0.12
		Control	3.97 ±0.11	4.61 ±0.03	5.01 ±0.12	3.18 ±0.11	3.10 ±0.20	3.02 ±0.08	2.70 ±0.02	2.94 ±0.05	3.26 ±0.10	3.89 ±0.19	4.29 ±0.20	4.05 ±0.14
5.	*Eucalyptus citriodora*	Effected	3.26 ±0.24	2.86 ±0.08	3.89 ±0.20	4.53 ±0.06	4.61 ±0.05	3.97 ±0.66	2.30 ±0.20	2.70 ±0.02	2.94 ±0.21	3.65 ±0.10	2.94 ±0.03	3.02 ±0.40
		Control	4.13 ±0.11	4.29 ±0.04	4.45 ±0.10	5.09 ±0.23	5.17 ±0.16	4.61 ±0.15	3.73 ±0.08	3.41 ±0.12	3.57 ±0.03	4.05 ±0.24	3.97 ±0.07	4.21 ±0.16
6.	*Ficus benghalensis*	Effected	2.70 ±0.05	3.26 ±0.54	3.89 ±0.12	4.05 ±0.05	2.46 ±0.02	2.22 ±0.21	1.66 ±0.10	2.14 ±0.20	3.26 ±0.07	3.73 ±0.24	2.94 ±0.03	3.10 ±0.30
		Control	3.81 ±0.03	3.89 ±0.21	4.13 ±0.03	4.77 ±0.11	4.13 ±0.23	3.26 ±0.12	3.10 ±0.66	3.73 ±0.12	3.97 ±0.14	4.29 ±0.20	4.21 ±0.15	3.97 ±0.12
7.	*Ficus religiosa*	Effected	3.97 ±0.55	3.73 ±0.21	3.65 ±0.08	4.29 ±0.20	3.81 ±0.03	2.86 ±0.25	2.14 ±0.15	1.90 ±0.08	3.26 ±0.14	2.70 ±0.03	2.76 ±0.04	2.70 ±0.20
		Control	3.97 ±0.25	4.61 ±0.11	5.01 ±0.1	5.25 ±0.14	4.69 ±0.04	3.02 ±0.12	2.78 ±0.05	2.38 ±0.11	4.05 ±0.21	4.13 ±0.05	3.89 ±0.10	3.81 ±0.12
8.	*Zizyphus jujuba*	Effected	3.65 ±0.08	3.26 ±0.10	3.81 ±0.14	4.29 ±0.12	3.10 ±0.21	2.86 ±0.04	2.54 ±0.23	2.46 ±0.15	2.94 ±0.02	2.94 ±0.25	3.02 ±0.21	3.73 ±0.30
		Control	4.29 ±0.11	4.37 ±0.20	4.45 ±0.03	4.93 ±0.15	4.13 ±0.15	4.05 ±0.24	3.18 ±0.02	3.02 ±0.20	3.18 ±0.11	3.81 ±0.30	4.13 ±0.16	4.21 ±0.15

Table 10.2 : Average amount of protein (mg g^{-1}) in the leaves of various trees of effected and control area

S. No.	Name of the species	Effected Control	AVERAGE AMOUNT OF PROTEIN (mg g^{-1})											
			Jul. 1988	Aug. 1988	Sep. 1988	Oct. 1988	Nov. 1988	Dec. 1988	Jan. 1989	Feb. 1989	Mar. 1989	Apr. 1989	May 1989	Jun. 1989
1.	*Azadirachta indica*	Effected	24.81 ±0.57	25.31 ±0.23	26.31 ±0.43	32.31 ±0.53	26.81 ±0.50	20.37 ±0.26	12.87 ±0.51	13.87 ±0.57	20.37 ±0.28	20.37 ±0.63	20.37 ±0.46	22.31 ±0.35
		Control	26.31 ±0.51	25.81 ±0.46	31.31 ±0.35	35.31 ±0.72	31.31 ±0.72	26.81 ±0.45	19.37 ±0.76	21.31 ±0.57	20.81 ±1.03	25.31 ±1.19	26.31 ±0.98	26.81 ±0.75
2.	*Bauhinia purpurea*	Effected	24.31 ±4.23	18.87 ±0.50	17.87 ±3.22	23.31 ±0.75	25.31 ±0.28	18.87 ±3.55	16.87 ±0.72	13.87 ±0.46	11.37 ±0.75	8.93 ±0.76	21.81 ±1.12	25.81 ±3.22
		Control	24.81 ±1.30	25.31 ±0.72	26.31 ±1.19	28.31 ±0.14	31.81 ±0.57	28.31 ±1.03	21.31 ±1.43	19.37 ±0.43	18.37 ±0.53	17.87 ±0.35	22.31 ±0.57	26.31 ±1.13
3.	*Cassia fistula*	Effected	18.87 ±0.23	24.31 ±0.43	22.31 ±1.21	27.31 ±0.63	25.31 ±0.57	14.87 ±0.72	8.93 ±1.02	17.37 ±1.03	21.81 ±0.72	18.87 ±0.80	23.81 ±0.20	19.37 ±1.30
		Control	22.81 ±0.31	24.81 ±0.98	26.81 ±0.43	30.81 ±0.35	29.81 ±0.45	21.31 ±1.12	18.37 ±1.19	20.37 ±0.57	22.81 ±0.28	23.81 ±0.46	24.81 ±1.12	24.31 ±0.57
4.	*Dalbergia sissoo*	Effected	20.37 ±0.69	25.31 ±0.72	27.31 ±0.51	13.37 ±1.87	13.87 ±0.46	12.87 ±0.76	10.37 ±0.80	15.87 ±0.75	15.87 ±0.31	21.81 ±0.72	22.31 ±0.53	16.87 ±0.50
		Control	24.81 ±0.50	28.81 ±0.45	31.31 ±0.50	19.87 ±0.79	19.37 ±0.60	18.87 ±0.53	16.87 ±0.57	18.37 ±0.45	20.37 ±0.50	24.31 ±1.03	26.81 ±0.60	25.31 ±0.79
5.	*Eucalyptus citriodora*	Effected	20.37 ±1.43	17.87 ±0.53	24.31 ±0.14	28.31 ±0.43	28.81 ±0.50	24.81 ±3.55	14.37 ±1.13	16.87 ±0.57	18.87 ±1.87	22.81 ±0.50	18.37 ±0.45	18.87 ±0.20
		Control	25.81 ±0.79	26.81 ±0.57	27.81 ±0.63	31.81 ±1.12	32.31 ±1.03	28.81 ±1.14	23.31 ±0.51	21.31 ±0.50	22.31 ±0.23	25.31 ±1.43	24.81 ±0.46	26.31 ±1.03
6.	*Ficus benghalensis*	Effected	16.87 ±0.35	20.37 ±3.22	24.31 ±0.50	25.31 ±0.50	15.37 ±0.57	13.87 ±0.60	10.37 ±0.63	13.37 ±1.13	20.37 ±0.45	23.31 ±1.12	18.37 ±0.75	19.37 ±1.70
		Control	23.81 ±0.23	24.31 ±1.87	25.81 ±0.75	29.81 ±0.72	25.81 ±1.12	20.37 ±0.76	19.37 ±3.55	23.31 ±0.98	24.81 ±0.76	26.81 ±1.19	26.31 ±1.14	24.81 ±0.50
7.	*Ficus religiosa*	Effected	24.81 ±2.21	23.31 ±1.70	22.81 ±1.43	26.81 ±1.14	23.81 ±0.75	17.87 ±1.54	13.37 ±0.69	11.87 ±0.28	20.37 ±0.76	16.87 ±0.23	17.37 ±0.45	16.87 ±1.13
		Control	24.81 ±1.54	28.81 ±0.80	31.31 ±0.21	32.81 ±0.76	29.31 ±0.28	18.87 ±0.57	17.37 ±0.35	14.87 ±0.98	25.31 ±4.12	25.81 ±0.50	24.31 ±0.63	23.81 ±0.76
8.	*Zizyphus jujuba*	Effected	22.81 ±0.28	20.37 ±0.50	23.81 ±0.69	26.81 ±0.57	19.37 ±1.12	17.87 ±0.57	15.87 ±1.87	15.37 ±0.69	18.37 ±0.57	18.37 ±1.54	18.87 ±0.60	23.31 ±1.70
		Control	26.81 ±0.80	27.31 ±0.14	27.81 ±0.23	30.81 ±0.69	25.81 ±1.03	25.31 ±1.54	19.87 ±0.57	18.87 ±1.87	19.87 ±0.72	23.81 ±1.70	25.81 ±1.03	26.31 ±0.69

The amount of total protein is based on the nitrogen contents therefore the amount of total nitrogen and total protein show the same type of trend.

The amount of total nitrogen and total protein was less in the trees growing in the polluted area than that of the control area throughout the investigation period i.e. from July 88 to June 89.

The seasonal variations in the amount of total nitrogen and total protein in the polluted and the control area trees is given in Tables 10.3, 10.4 and Figs. 10.1 - 10.16.

In affected trees of *Azadirachta indica, Cassia fistula, Dalbergia sissoo, Ficus benghalensis, Ficus religiosa* and *Zizyphus jujuba,* the amount of total nitrogen and total protein showed seasonal variations. The amounts were highest during the rainy season, the same decreased during winter season and increased in the summer season.

In *Bauhinia purpurea* and *Eucalyptus citriodora* the amounts of total nitrogen and total protein were also highest during the rainy season, then a decrease was noticed during the winter as well as the summer season. The lowest amounts were recorded in the summer season.

The amount of total nitrogen and total protein in all the species growing in the polluted area during rainy, winter and summer seasons showed lower amounts in comparison to that of the control.

The amount of total nitrogen and total protein in affected trees of *Ficus benghalensis* and *Zizyphus jujuba* were significantly lower than that of the control during rainy, winter and summer seasons.

In *Azadirachta indica* and *Cassia fistula* amounts of total nitrogen and total protein were significantly lower during summer season whereas during rainy and winter season though the amounts were lower but the differences were not significant when compared to that of the control.

In *Eucalyptus citriodora* and *Ficus religiosa* the significant differences in the amounts of total nitrogen and total protein were found in the rainy and summer seasons whereas during winter the differences were not significant.

Table 10.3 : Seasonal variation in the average amount of Nitrogen ($mg\ g^{-1}$) in the leaves of Control and effected area trees

S. No.	*Name of species*	*Seasons*	*Control*	*Effected*	*Calculated t value*
1.	*Azadirachta indica*	Rainy	4.75	4.35	1.42
		Winter	3.95	2.95	2.10
		Summer	3.97	3.33*	4.0
2.	*Bauhinia purpurea*	Rainy	4.19	3.37*	4.10
		Winter	4.03	2.99*	2.44
		Summer	3.39	2.71	1.34
3.	*Cassia fistula*	Rainy	4.21	3.71	1.81
		Winter	3.59	2.66	1.95
		Summer	3.83	3.35*	3.55
4.	*Dalbergia sissoo*	Rainy	4.19	3.45	1.66
		Winter	2.94	2.12*	5.85
		Summer	3.87	3.19*	3.23
5.	*Eucalyptus citriodora*	Rainy	4.49	3.63*	2.91
		Winter	4.23	3.39	1.78
		Summer	3.95	3.13*	5.46
6.	*Ficus benghalensis*	Rainy	4.15	3.47*	2.56
		Winter	3.55	2.12*	7.30
		Summer	4.11	3.25*	6.61
7.	*Ficus religiosa*	Rainy	4.71	3.91*	3.47
		Winter	3.21	2.67	1.16
		Summer	3.97	2.86*	10.57
8.	*Zizyphus jujuba*	Rainy	4.51	3.75*	4.22
		Winter	3.59	2.74*	3.77
		Summer	3.83	3.15*	3.23

* Significant at 5% level for 6 degrees of freedom.

Table 10.4 : Seasonal variation in the average amount of Protein ($mg\ g^{-1}$) in the leaves of Control and effected area trees

S. No.	*Name of species*	*Seasons*	*Control*	*Effected*	*Calculated t value*
1.	*Azadirachta indica*	Rainy	29.68	26.93	1.42
		Winter	24.70	18.48	2.10
		Summer	24.81	20.87*	4.00
2.	*Bauhinia purpurea*	Rainy	26.18	21.09*	4.10
		Winter	25.20	18.73*	2.44
		Summer	21.21	16.84	1.34
3.	*Cassia fistula*	Rainy	26.31	23.20	1.81
		Winter	22.46	16.62	1.95
		Summer	23.93	20.96*	3.55
4.	*Dalbergia sissoo*	Rainy	26.20	21.59	1.66
		Winter	18.37	13.24*	5.85
		Summer	24.20	19.96*	3.23
5.	*Eucalyptus citriodora*	Rainy	28.06	22.71*	2.91
		Winter	26.43	21.21	1.78
		Summer	24.68	19.60*	5.46
6.	*Ficus benghalensis*	Rainy	25.93	21.71*	2.56
		Winter	22.21	13.24*	7.30
		Summer	25.68	20.35*	6.61
7.	*Ficus religiosa*	Rainy	29.43	24.43*	3.47
		Winter	20.06	16.68	1.16
		Summer	24.81	17.87*	10.57
8.	*Zizyphus jujuba*	Rainy	28.18	23.45*	4.22
		Winter	22.43	15.72*	3.77
		Summer	23.93	19.68*	3.23

* Significant at 5% level for 6 degrees of freedom.

AZADIRACHTA INDICA

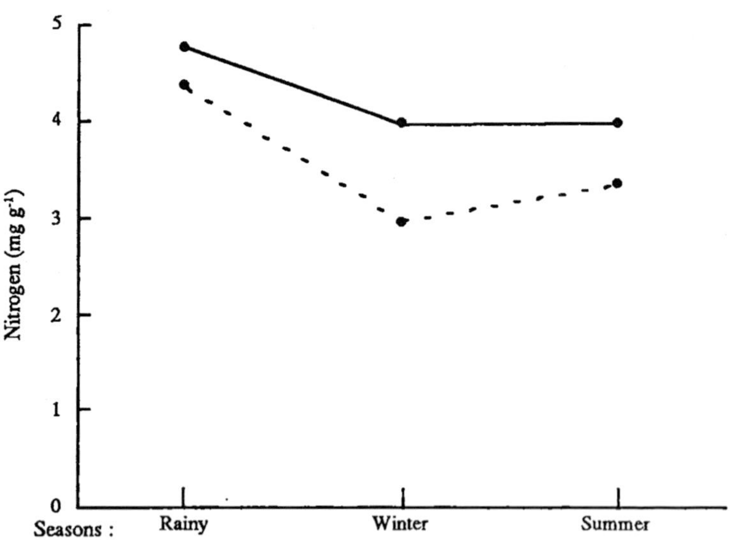

Fig. 10.1

BAUHINIA PURPUREA

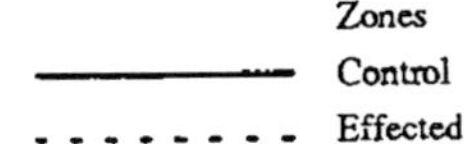

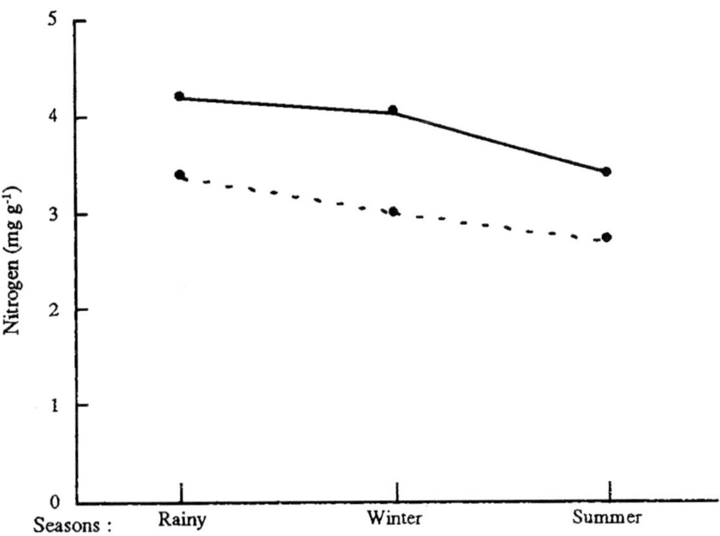

Fig. 10.2

AZADIRACHTA INDICA

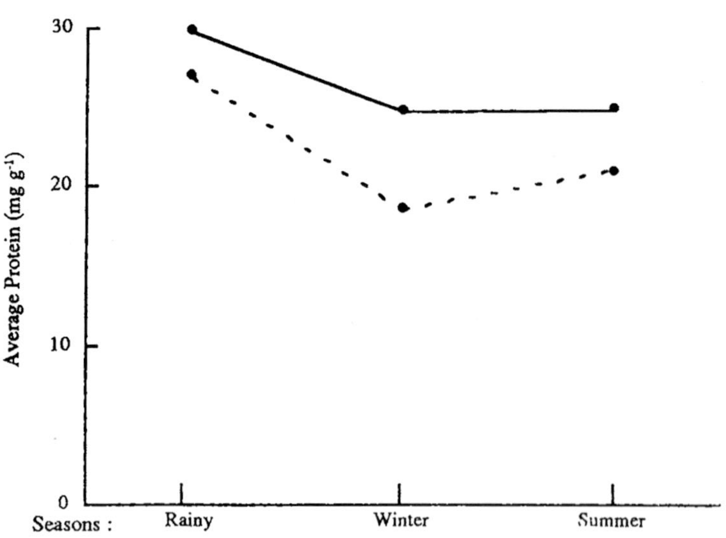

Fig. 10.3

BAUHINIA PURPUREA

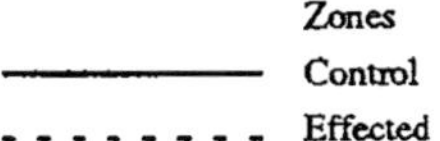

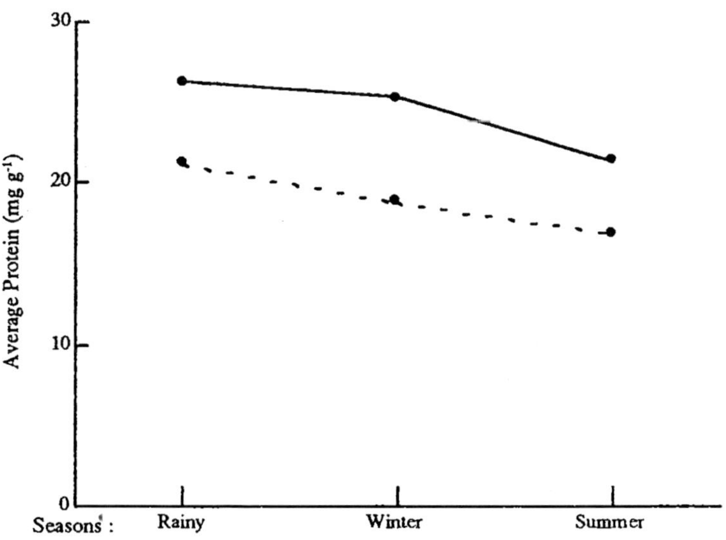

Fig. 10.4

CASSIA FISTULA

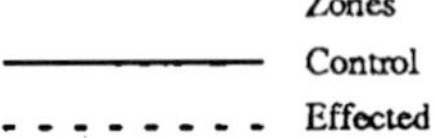

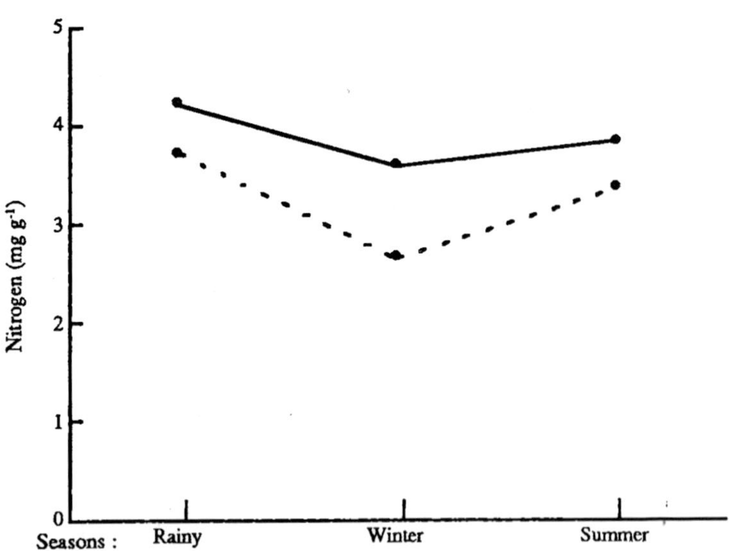

Fig. 10.5

DALBERGIA SISSOO

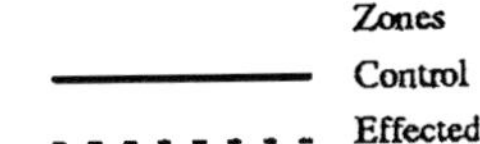

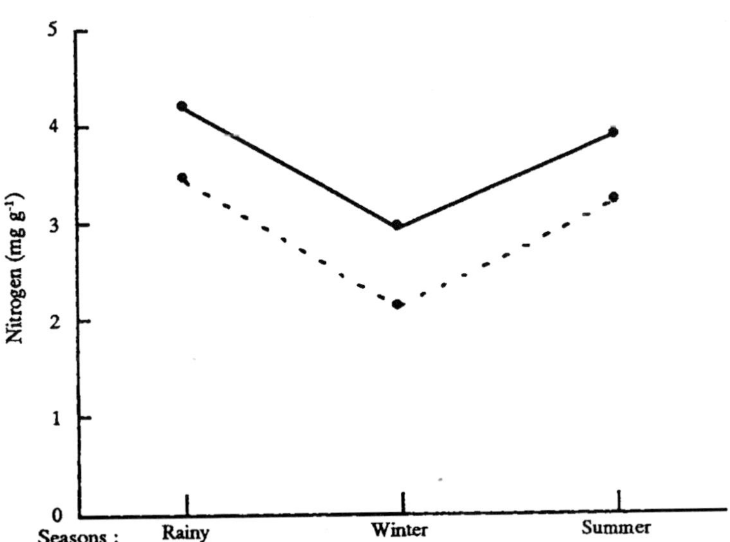

Fig. 10.6

CASSIA FISTULA

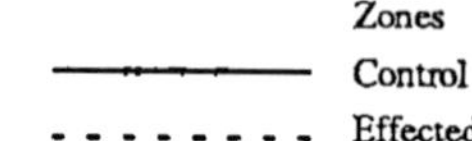

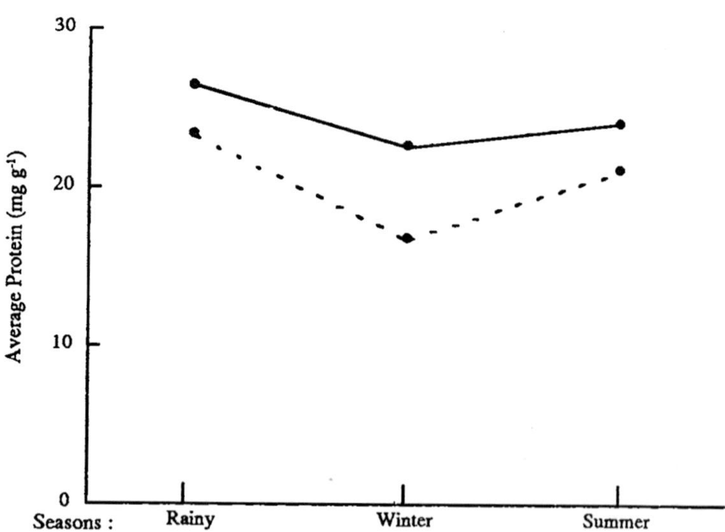

Fig. 10.7

DALBERGIA SISSOO

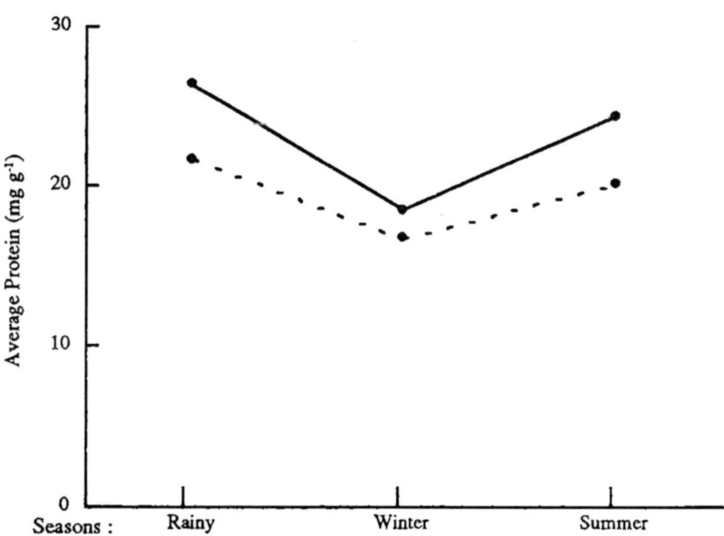
Zones
Control
Effected
30
20
10
0
Average Protein (mg g-1)
Seasons :
Rainy
Winter
Summer

Fig. 10.8

EUCALYPTUS CITRIODORA

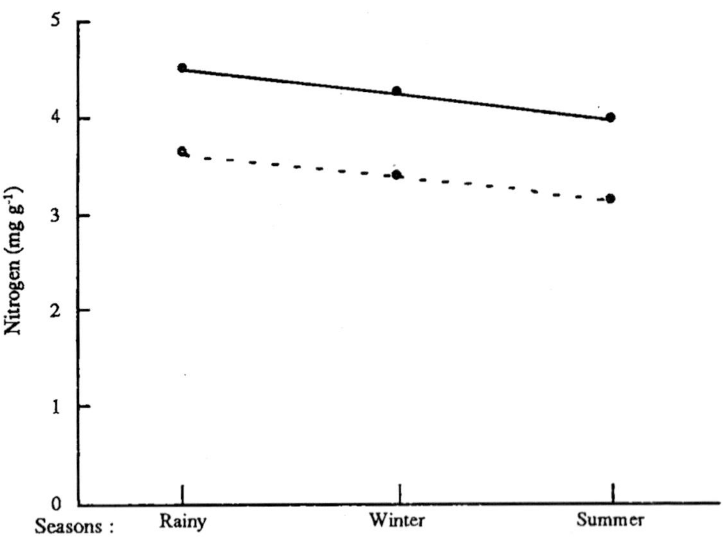

Fig. 10.9

FICUS BENGHALENSIS

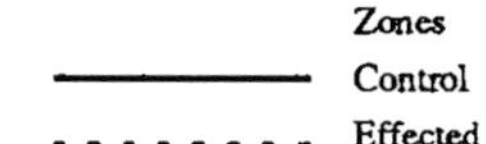

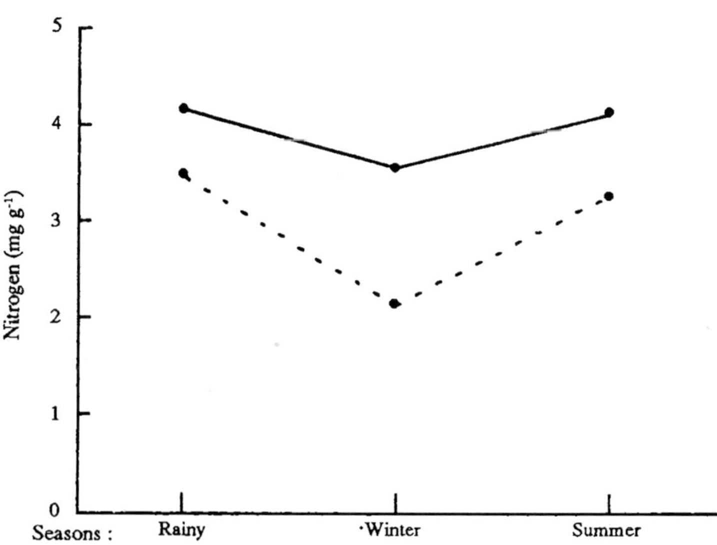

Fig. 10.10

EUCALYPTUS CITRIODORA

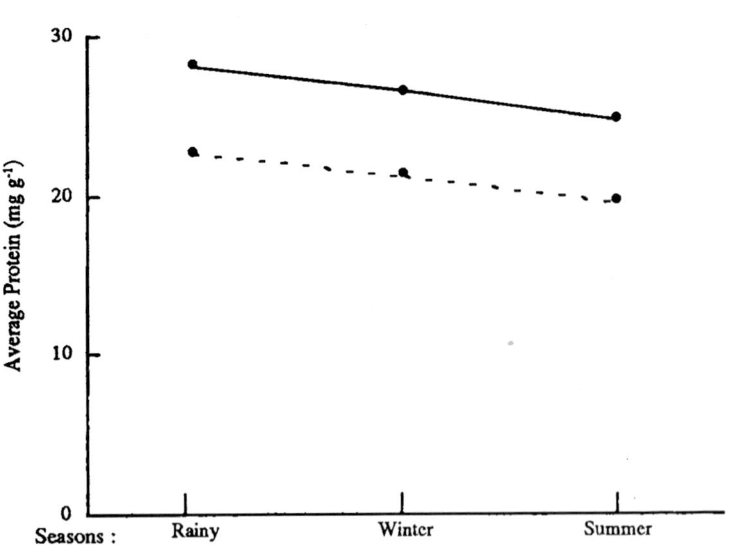

Fig. 10.11

FICUS BENGHALENSIS

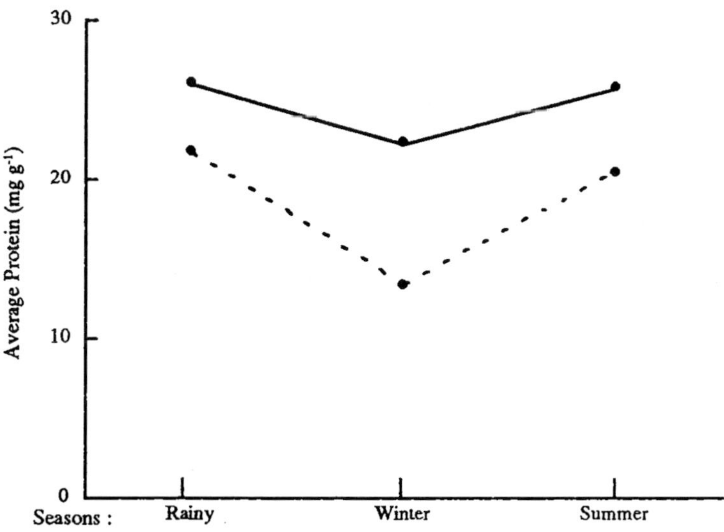
Zones
Control
Effected
30
20
10
0
Average Protein (mg g-1)
Seasons :
Rainy
Winter
Summer

Fig. 10.12

FICUS RELIGIOSA

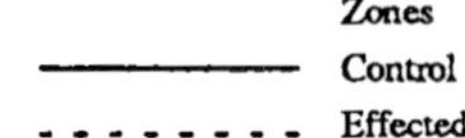

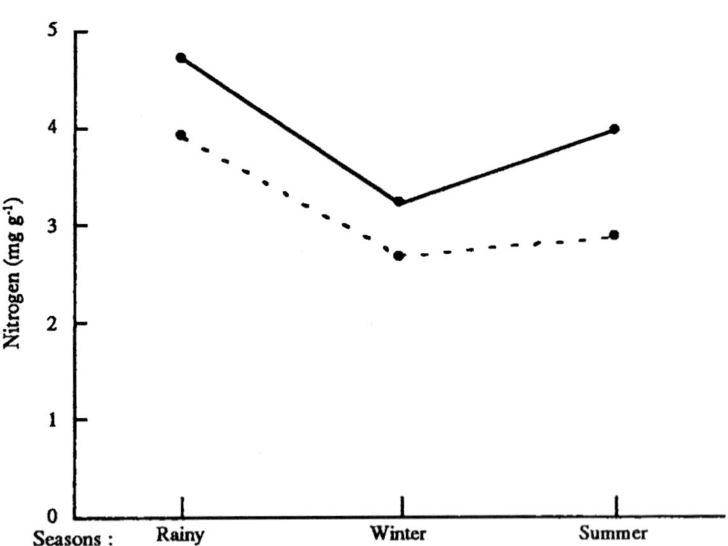

Fig. 10.13

ZIZYPHUS JUJUBA

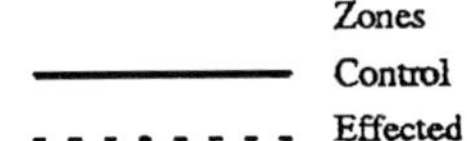

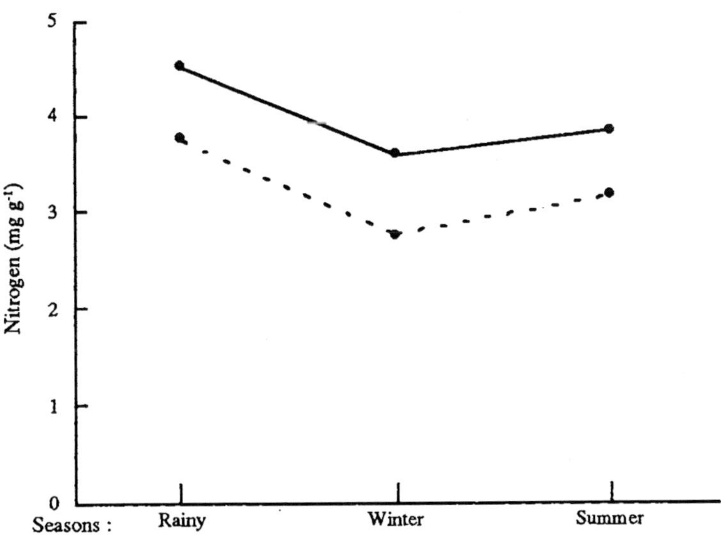

Fig. 10.14

FICUS RELIGIOSA

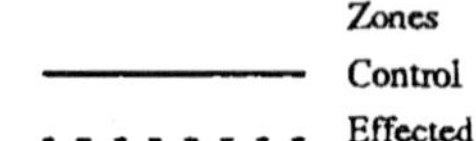

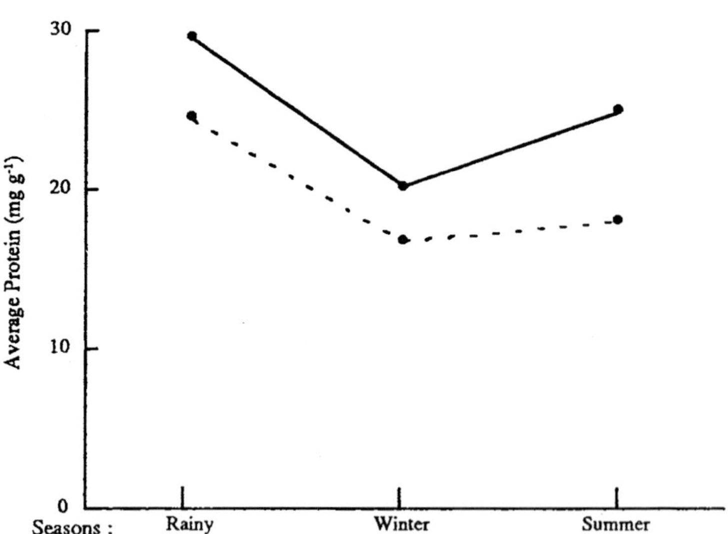

Fig. 10.15

ZIZYPHUS JUJUBA

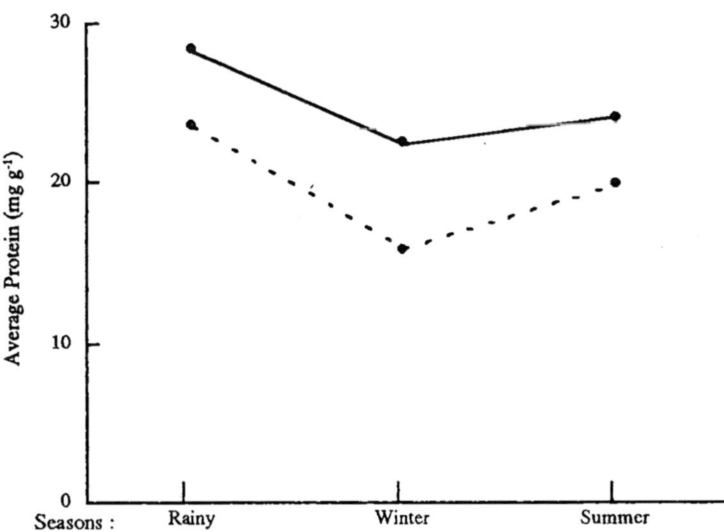

Fig. 10.16

Bauhinia purpurea growing in the polluted area, shows significantly lower amounts of total nitrogen and total protein during rainy and winter season whereas *Dalbergia sissoo* growing in the polluted area showed significantly lower amounts during winter and summer seasons in comparison to that of the control.

DISCUSSION

In the present study the amount of total nitrogen and total protein an *Azadirachta indica, Bauhinia purpurea, Cassia fistula, Dalbergia sissoo, Eucalyptus citriodora, Ficus benghalensis, Ficus religiosa* and *Zizyphus jujuba* exposed to the continuous emissions from HEG factory decreased from July 1988 to June 1989 (period of investigation) in comparison to that of the control. The decrease in the amount of nitrogen due to various pollutants has been reported by number of workers. Sankhla *et al.* (1982) working on *Calotropis procera* and *Melia azadirachta* reported lowering in nitrogen and phosphorus content due to coal dust deposition on the leaves. Sashikala *et al.* (1988 a & b) reported decreased amount of nitrogen in soybean at 5 ppm NO_2 exposure. Sharma and Rao (1985) working on *Phaseolus aureus* reported decreased amount of nitrogen with increased cumulative dose of SO_2 fumigation. Hallgren and Huss (1975), Prasad *et al.* (1983) working on SO_2 exposed plants, Rao and Prasad (1983) working on effect of automobile exhaust, Shrivastava and Ormord (1986) on NO_2 and nitrate nutrition, reported that the nitrogen was lowered. The decreased amount of nitrogen and protein also corroborated with the findings of Dikshit (1991) who worked on the exposure of toxic gas on *Azadirachta indica, Mangifera indica, Polyalathia longifolia, Psidium guajava, Zizyphus jujuba, Cleodendron inerme* and *Cynodon dactylon.*

Farooq *et al.* (1985) reported that SO_2 exposure decreased the protein contents in *Holoptelea integrifolia.* Yunus *et al.* (1985) reported decreased proteins in *Calendula* and *Dahlia* due to CO_2 exposure. Murray and Wilson (1988) reported that HF + SO_2 decreased protein concentration in *Eucalyptus marginata.* Saroja and Bose (1982) working on effect of methyl parathion insecticide on *Chlorella protothecoides,* Prasad *et al.* (1984), Agrawal *et al.* (1987), Kumar and Singh (1988) working on exposure of SO_2 on *Triticum aestivum, Oryza sativa* L. var. Jaya and *Mangifera indica, Psidium guajava* respectively, reported decreased amounts of protein. Agrawal *et al.* (1985) found in *Vicia faba* that the protein contents in the leaves decreased due to the effect of O_3.

The decreased amount of nitrogen and protein due to the emissions might be the result of decreased photosynthesis (Siz and Swanson, 1974; Constantinidou and Kozlowski, 1979) or inhibition of protein synthesis or enhanced protein degradation (Robe and Kreeb, 1980). Similar observations have been made by Tomlinson and Rich (1967), Ting and Mukerji (1971) and Dikshit (1991).

The amount of total nitrogen and total protein during the rainy, winter and summer seasons in all the eight species growing in the polluted area under investigation were less than that of the control. The total nitrogen and total protein were highest during the rainy season and lowest during winter whereas the amounts increased towards summer season in all the species except *Eucalyptus citriodora* and *Bauhinia Purpurea.*

The higher amount of total nitrogen and total protein during rainy season might be due to the fact that during vegetative growth period with plenty of water available in the rainy season, the absorption of minerals might have increased (Prasad *et al.*, 1983; Dikshit, 1991) consequently increasing the amount of total nitrogen and total protein. It seems that the dust fall and other gaseous emissions emitted from the HEG factory were diluted and washed away during rainy scason resulting in the increased amount of nitrogen and protein during this period. In the winter the amount of dust fall in the polluted area being more has resulted in the decreased protein synthesis. During the summer season which corresponds to the formation of fruits due to accumulation of proteins the amount of nitrogen and protein increased (Prasad and Rao, 1979).

REFERENCES

1. Agrawal, M., P.K. Nandi and D.N. Rao 1985. Responses of *Vicia faba* plants to O_3 pollution. Ind. J. of Environ. Hlth., 27(4) : 319-329.

2. Agrawal, S.B., M. Agrawal, P.K. Nandi and D.N. Rao 1987. Proc. 73rd Ind. Sci. Cong. Part III, Abstracts 353, p. 242.

3. Constantinidou, H.A. and T.T. Kozlowski 1979. Effects of sulphur dioxide and ozone on *Ulmus americana* seedlings II. Carbohydrates, Protein and Lipids. Can. J. Bot., 57 : 176-184.

4. Dikshit, S. 1991. Effect of toxic gas leakage on the pollen sterility, ascorbic acid and nitrogen and protein contents of some terrestrial plants of Bhopal. Ph.D. Thesis, Barkatullah University, Bhopal.

5. Farooq Mohd., Akbar Masood and M.U. Beg 1985. Effect of acute exposure of sulphur dioxide on the metabolism of *Holoptelea integrifolia* plants. Environ. Pollut. (Series A), 39 : 197-205.

6. Hallgren, J.E. and K. Huss 1975. Effects of SO_2 on photosynthesis and nitrogen fixation. Physiol. Plant, 34 : 171-176.

7. Jackson, M.L. 1962. Soil chemical analysis. Prentice Hall Inc., New Jersery, USA.

8. Kumar, N. and V. Singh 1988. Sensitivity of *Mangifera indica* and *Psidium guajava* plants to SO_2 pollution. Natl. Acad. Sci. Lett., 11(6) : 167-169.

9. Murray Frank, Susan Wilson 1988. Effects of sulphur dioxide hydrogen fluoride and their combination on three Eucalyptus species. Environ. Pollut., 265-279.

10. Prasad, B.J. and D.N. Rao 1979. Morphological responses of wheat plants to NO_2 exposures. Proc. Indian Natn. Sci. Acad., B. 45(2) : 142-146.

11. Prasad, B.J., D.N. Rao and U.R. Singh 1983. Metabolic responses of SO_2 exposed soybean and wheat plants. In : "Air Pollution : Problem and Perspectives". Ed. D.N. Rao. T.P. Sharma and U.R. Singh, Publ. EPCO, Paryavaran Parisar, Bhopal, pp. 223-245.

12. Rao, D.C. and B.J. Prasad 1983. Phytotoxicity of Gasoline Vapour. In "Air Pollution Problems and Perspective". Edts. D.N. Rao, T.P. Sharma and U.R. Singh. Publ. EPCO, Paryavaran Parisar, Bhopal, pp. 209-221.

13. Robe, R. and K.H. Kreeb 1980. Effects of SO_2 upon enzyme activity in plant leaves. Inter. J. Plant Physoil., 97 : 215-226.

14. Sankhla, S.K., M. Yusuf, P.P. Paliwal and L.N. Vyas 1982. Effect of coal dust pollution on some plant species growing at Bajaj Nagar Udaipur. Advances in Environmental Research, pp. 26-29.

15. Saroja G. and S. Bose 1982. Effects of methyl parathion on the growth, cell size, pigment and protein content of *Chlorella protothecoides*. Environ. Pollut. (Series A) 27(4) : 297-308.

16. Sashikala Sabaratnam, Gian Gupta and Charles Mulchi 1988a. Effects of nitrogen dioxide on leaf chlorophyll and nitrogen content of soybean. Environmental Pollution, 51 : 113-120.

17. Sashikala Sabaratnam and Gian Gupta 1988b. Effects of nitrogen dioxide on biochemical and physiological characteristics of soybean. Environmental Pollution, 55 : 149-158.

18. Sharma, H.C. and D.N. Rao 1985. Responses of *Phaseolus aureus* plants to SO_2 and HF pollutants. Perspectives in Environmental Botany, 131-150.

19. Shrivastava, H.S. and D.P. Ormord 1986. Effects of NO_2 and nitrate nutrition on modulation on nitrogenase activity, growth and nitrogen content of Bean plants. Plant Physiol. 81 : 737-741.

20. Sij. J.W. and C.A. Swanson 1974. Short term kinetic studies on the inhibition of Photosynthesis by sulphur dioxide. J. Environ. Qual., 3 : 103-107.

21. Ting, I.P. and S.K. Mukerji 1971. Leaf ontogeny as a factor in susceptibility to ozone; amino acids and carbohydrate changes during expansion. Am. J. Bot. 58 : 497-504.

22. Tomlinson H. and S. Rich 1967. Metabolic changes in free amino acids of bean leaves exposed to ozone. Phytopathol, 57 : 972-974.

23. Yunus, M., S.N. Singh, K. Shrivastava, K. Kulshrestha and K.J. Ahmad 1985. Relative sensitivity of *Calendula* and *Dahlia* to SO_2. Perspectives in Environmental Botany, 271-284.

EFFECT OF ORDNANCE FACTORY EFFLUENT ON SEED GERMINATION AND EARLY GROWTH PERFORMANCE OF PEA SEEDS

R.K. Srivastava, Akhilesh Ayachi, Monica Pandya, Bhavna Khare

ABSTRACT

Effect of ordnance factory effluent on seed germination and early growth performance of pea seeds was studied. It was found that the ordnance factory effluent was highly deleterious for the germination and early growth performance of seeds and as the concentration of effluent increases the deleterious effect also increases thereby showing positive correlation.

INTRODUCTION

Water pollution due to industrial effluents have received the considerable attention these days and it is increasing day by day. The effects of these industrial effluents on the growth and physiology of various organisms have been investigated by various workers (Rajaram & Oblisamy, 1979; Spulnik, 1940; McCermic, 1959; Mathur *et al.*, 1984; Srivastava, 1991; Mathur *et al.*, 1987; Srivastava & Mathur, 1987; Goel and Kulkarni, 1994; Dubey and Dwivedi, 1987).

The present study attempts to find out the effects of industrial effluent of ordnance factory, Jabalpur on germination and early growth performance of pea seeds.

MATERIAL AND METHODS

The various Chemicals are utilized for manufacture and rust proof coating of ammunition and effluent is discharged. The samples are collected from the outlet and were analysed for various physio-chemical parameters as per standard methods (APHA 1971).

Healthy Pea seeds of previous season were procured. Different dilution of the effluents were made. (10, 30, 50, 70 & 100%) and seeds were soaked separately in each of these solutions for an hour. In control the seeds were pre-soaked in distilled water for the same duration of time. 15 ml. of each sol. was used to soak the filter paper in each of the sterilized Petri-dishes of 20 cm. dia. on which 25 seeds were spread per Petri-dish. Three replicates in each treatment were put for the treatment. A control was seen simultaneously with distilled water.

The germinated seeds were counted after 24 hours and daily progress in germination was recorded for 5 days. After the 5th day the length of radicle and plumule was recorded for 20 randomly selected seedlings to study the early growth performance. The experiments were carried out in the laboratory conditions of bright light at 22-26°C and 40-70% RH.

RESULT AND DISCUSSION

The Table 11.1 shows the chemical characteristics of O.F.K. effluent. These are very high concentration of heavy metals. The Table 11.2 shows the percent germination of seeds in different dilution of factory effluent.

The increasing concentration of (10, 30, 50, 70, 100%) effluent have deleterious effects as observation from the Table 11.2. The most affected is the 100% effluent treated seeds where only 29% germination is observed after max of five days, the least affected was 10% Conc. effluent seeds where 60% germination was observed. In control whereas it was 84%. No germination was observed in 70 & 100 and effluents on the first day.

Table 11.3 shows the mean radicle length of seeds recorded for different dilution of effluents. In control it is max 2.5 cm. as compared to toust in 100% effluent and it is 3 cm.

Table 11.1 : Physico-Chemical Characteristics of Ordnance Factory effluent

S. No.	Characteristics	in mg/1.
1.	Suspended solids	100 - 300
2.	T.D.S.	125
3.	Oil and Grease	8 - 15
4.	BOD 5 day 20°C	30 - 70
5.	COD	240 - 250
6.	Chromium	0.01 - 0.1
7.	Copper	0.887 - 3.0
8.	Cadmium	0.3 - 1.1
9.	Alkalinity	160
10.	pH	6 - 9

Table 11.2 : Percent germination of Pea seeds treated with different dilution of effluents

Core of Effluent	Days	/	%	germination	
	1	2	3	4	5
10	20	30	40	60	60
30	20	30	35	55	55
50	15	25	38	48	48
70	-	15	20	35	35
100	-	10	18	29	29
Control	30	50	68	84	84

Table 11.3 : Length of Radicle of Pea seeds treated with diff. dilution of effluents

Concentration	Length (mean cm.)
10	1.7
30	1.4
50	1.1
70	0.6
100	0.3
Control	2.5

The mean length of radicle varies from 0.3 cm. to 2.5 cm., (0.3 in 100%, 0.6 in 70%. 1.1 in 50%, 1.4 in 30% and 1.7 in 10%).

The toxic effluent which comprises heavy metals in higher concentrations is probably causing the retardation in germination as well as early growth performance. A some what similar observation with Lead was noted by Mukherji and Maitra (1976) in their studies on rice and onion. Toxic effect of Lead was observed on growth and metabolism of germinating rice seeds and on mitosis of onion root tip cells.

Purohit and Ametel (1975) have reported inhibition of the root growth of onion bulbs took place in only 25, 50 and 75 ppm sodium arsenite sol. Mathur *et al.*, (1987) and Srivastava (1991) also observed similar results. Mathur *et al.*, (1989) found that Cadmium and Chromium metals have toxic effects on germination and early growth performance of onion seeds. Srivastava (1989) showed that paper mill effluent and Chlor alkal plant effluent were toxic to Radish and Onion seeds as far as germination and early growth performance studies are concerned.

ACKNOWLEDGEMENT

The authors are thankful to Dr. S.S. Ayachi, Prof. & Head, Deptt. of Botany and Environmental Science, Govt. Autonomous Science College and the Principal, Govt. Autonomous Science College, Jabalpur for providing necessary facilities for the work.

REFERENCES

1. Dubey, R.C. and Dwivedi, R.S. 1987. Effect of heavy metals on seed germination and seedling growth of Soybean. Nat. Acad. Sci., Lett. 10(4), 121-123.

2. Goel, P.K. and Kulkarni, S.M. 1994. Effects of Sugar factory waste on germination of Gram seeds. Jour. of Env. & Poll., Vol 1(1), 35-43.

3. Mathur, K.C., Srivastava R.K. and Chaudhary Kanchana 1984. Some studies on the toxicity of popular industrial metals on the germination of onion seeds. Paper presented at National Seminar EPCO, Bhopal, pp. 50-51.

4. Mathur, K.C., Srivastava, R.K., and Kanchan Chaudhary 1987. Effect of Cadmium and Chromium metals on germination and early growth performance of *Alium cepa* seeds. Proc. Nat. Acad. Sci. India, 57, (B) II.

5. McCermic, C.L. 1959. Effect of paper mill waste water on crops and soil. Bull. La. Agric. : Exp. stn. 25 : 302-310.

6. Mukherjee S. and Maitra P. 1976. Ind. I Exp. Biol. 14 : 519.

7. Purohit, S.S. and Ameta S.C. 1975. Veg. par. Anu. Pat. 18 : 37.

8. Rajaram G. and Oblisamy G. 1979. Effect of paper factory effluent on soil and crop plants. Ind. J. Environ. Hlth. 21 : 120-130.

9. Spulnik, J.B. 1940. Effect of Waste sulphite liquor on soil properties and plant growth. Soil Sciences. 49 pp. 320-327.

10. Srivastava, R.K. 1991. Effect of Paper mill effluent on seed germination and early growth performance of Radish and Onion. J. Ecotoxicol. Environ. Monit. 1(1). 13-18.

AN ANALYSIS OF THROUGHFALL AND STEMFLOW IN AN ARTIFICIAL PLANTATION SITE

A.K. Girolkar

ABSTRACT

The chapter deals with rainfall, throughfall and stemflow measured in a *Terminalia tomentosa* plantation site of Raigarh forest division of Madhya Pradesh, India.

The observations made from June 15 to September 30, 1985, suggest that:

1. Throughfall was variable from point to point and averaged 62.69% of the total rainfall (688.75 m.m.).

2. Stemflow averaged 37.53% of the total rainfall.

INTRODUCTION

Fater cycling studies in the forest ecosystem is lesser studied part. Water moves materials within ecosystem and therefore understanding of water flow is an important aspect of ecosystem studies.

Rainfall reaches the forest floor in two major ways, freely falling or dripping (throughfall) and running down the trunks of trees (stemflow) Raich (1983). Havely and Partic (1965) state that a number of investigations have been carried out on the amount of throughfall and stemflow in temperate regions while, Comparatively little has been reported from tropical areas (Nye 1961; Sollings and Drewry 1970; Clements and Colon 1975; Golley et. al. 1975; Jordan 1978; Raich 1983; Yadav and Mishra 1985).

STUDY SITES

The study was conducted at *Terminalia tomentosa* plantation site Lakha, lies in the Urdana forest range of Raigarh forest division of Madhya Pradesh, India.

The climate is typically monsoonic and the year is distinctly divisible in to Rainy (mid June to September), mild winter (mid October to February) and hot summer (March to mid June) seasons. The average annual rainfall in the area is about 1639 m.m. where 90% of the rainfall is received in rainy months.

MATERIAL AND METHODS

Rain gauzes were placed for the collection of precipitation in the open (rainfall) and beneath the tree canopies (throughfall). In all 15 (fifteen) throughfall collectors were placed randomly.

Stemflow was collected by collars around the trees of different circumferences, collars were made by cutting Bicycle tube which completely encircled each tree trunk. The collars were made water tight with the help Fevicol (an adhesive). The water collected in the collar was drawn in to plastic collection bags through a plastic tubing. Throughfall and stemflow were measured daily.

RESULT

The site received 42.02% of the average rainfall (1639 m.m.) by the 30 September, 1985. Throughfall averaged 62.96% of the total rainfall received at the site for the period from June 15, to September 30, 1985 (Table 12.1). It was variable not only from rainyday to rainyday but also from point to point within the site.

The average volume of stemflow collected by each tree for each m.m. of rain falling in the open was 0.375 m.m. (37.53%), Table 12.1. It was observed to be variable with the trees of different girth classes.

ACKNOWLEDGEMENT

The author wishes to thank the Madhya Pradesh Council of Science and Technology (MAPCOST), Bhopal for financial assistance.

Table 12.1 : Datewise rainfall, throughfall and Stemflow

Month Date	Rain-fall mm.	Throughfall m.m.			Stemflow (in Lt.) Circ. of trees				mean
		Min.	Max.	Mean of 5	<35 cm.	>35 cm.	>40 cm.	>45 cm.	
June 16	10.00	2.00	16.00	7.00	4.96	3.46	3.02	3.09	3.63
23	2.25	-	2.78	1.85	2.70	1.30	1.30	0.25	1.40
27	8.50	3.00	17.50	3.90	5.60	1.75	0.25	1.65	2.31
28	1.00	0.20	0.90	0.40	0.38	0.11	0.70	0.10	0.32
30	44.00	5.00	40.00	26.00	21.83	15.22	8.95	13.50	14.90
July 01	24.00	7.00	21.00	13.05	10.35	8.75	6.46	7.46	8.22
03	2.25	0.75	2.75	1.70	0.95	0.20	0.18	0.00	0.33
04	6.75	2.50	12.25	4.75	7.21	2.70	1.38	1.35	3.16
05	40.75	10.75	47.00	25.00	43.53	16.30	8.85	10.25	19.72
06	5.00	2.00	9.25	2.00	4.50	2.10	0.85	1.60	2.26
07	4.00	1.50	7.00	3.00	4.30	1.57	0.88	2.30	2.26
09	8.25	2.00	13.50	3.50	0.75	4.45	2.22	4.40	4.95
11	26.75	3.00	33.00	15.00	28.77	14.30	5.75	14.21	15.67
13	1.00	0.00	2.50	0.60	1.30	0.25	0.88	0.30	0.58
15	1.50	0.00	6.50	1.00	2.00	0.70	0.37	0.45	0.95
16	18.25	5.50	28.75	10.50	11.60	10.45	3.37	10.50	8.89
17	4.00	1.00	7.00	3.00	3.40	1.70	0.40	0.90	1.62
20	3.25	1.50	5.25	1.18	2.90	1.12	0.40	0.75	1.29
22	2.00	0.50	3.75	1.25	1.95	1.00	0.30	0.65	0.99
24	9.50	2.50	15.25	5.25	6.25	4.90	1.60	4.38	4.28
25	0.50	0.00	0.75	0.20	0.15	0.00	0.00	0.10	0.06
26	20.00	8.00	42.00	17.00	16.90	13.41	5.47	11.76	11.88
27	14.50	3.50	31.00	8.65	6.35	4.35	1.85	4.26	4.20
29, 30	45.50	7.50	78.00	35.25	10.93	13.40	10.40	13.37	14.27
Aug. 02	36.25	9.00	97.50	19.50	10.90	7.70	10.70	10.80	10.03
03	8.50	1.50	18.50	5.00	7.85	1.45	7.45	6.95	5.93
04	1.00	0.00	2.25	0.50	0.35	0.45	0.48	0.55	0.46
05	3.50	0.00	10.50	2.00	1.00	0.88	1.65	1.80	1.35

Month Date	*Rain-fall mm.*	*Throughfall m.m.*			*Stemflow (in Lt.) Circ. of trees*				*mean*
		Min.	*Max.*	*Mean of 5*	*<35 cm.*	*>35 cm.*	*>40 cm.*	*>45 cm.*	
06	6.0	1.00	14.25	4.00	5.45	0.96	2.85	1.65	2.73
07	24.00	5.50	58.50	16.00	21.80	4.70	11.40	10.00	11.97
08	28.00	8.00	59.00	18.00	25.43	6.45	13.40	11.67	14.24
09	21.50	10.50	38.50	19.50	19.53	6.45	10.50	10.40	11.72
14	17.50	6.75	30.50	10.00	15.90	3.75	7.60	5.65	8.23
15	7.75	2.00	7.00	5.25	1.25	0.00	3.35	0.00	1.15
16	16.00	3.00	28.00	9.00	2.80	0.00	7.75	0.00	2.49
18	0.50	3.50	11.50	6.00	3.20	1.80	4.55	0.00	2.39
19	29.00	11.00	46.75	18.50	4.55	3.95	12.60	0.00	5.27
20	14.25	6.50	13.00	10.00	2.15	1.25	3.98	1.45	2.11
23	6.50	1.00	2.25	3.50	0.45	0.25	0.45	0.38	0.38
24	13.75	4.00	10.50	7.75	2.25	1.40	2.35	1.55	1.89
25	6.00	1.00	12.50	4.25	0.75	1.30	3.30	0.95	1.57
26	12.50	3.00	21.50	8.35	6.50	2.25	7.35	1.30	4.35
28	7.50	1.50	13.50	4.25	2.90	1.40	4.30	0.45	2.26
29	9.50	3.15	7.00	7.25	2.75	1.90	6.45	0.60	2.93
30	1.00	0.00	1.50	0.70	0.20	0.15	0.25	0.00	0.16
Sept. 02	8.00	1.50	19.50	3.65	3.55	1.85	3.45	2.35	2.77
03	13.75	3.20	10.00	9.25	4.15	3.25	7.35	2.75	4.38
04	8.75	2.50	10.50	7.00	2.75	1.65	3.55	1.95	2.45
07	22.50	9.75	44.00	14.00	4.55	7.40	10.35	4.35	6.66
10, 11, 12	2.00	0.00	5.50	0.70	1.45	0.85	1.35	0.75	1.10
15	14.00	4.00	34.50	7.00	3.65	4.65	8.75	0.30	4.38
16	1.00	0.00	3.50	0.70	0.65	0.30	0.50	0.25	0.42
21	15.50	5.25	27.00	9.25	4.55	2.60	8.55	2.65	4.50
23	2.00	0.00	5.50	1.50	0.95	0.55	0.95	0.57	0.76
25	12.25	3.25	22.00	8.25	4.40	4.33	10.00	2.57	5.32
Total	688.75	-	-	433.65	390.05	141.34	310.35	192.28	258.50

REFERENCES

1. Clements, R.G. and J.A. Colon, 1975: The rainfall interception process and mineral cycling in a montane rain forest in eastern Puerto Rico. In *Mineral cycling in South-Eastern Ecosystems.* U.S. Energy Research and Development Administration, Washington D.C.

2. Golley, G.B. et. al. 1975: *Mineral cycling in a Tropical Moist Forest Ecosystem,* University of Georgia Press Athens.

3. Helvey, J.O. & J.H. Patric, 1965: Canopy and litter interception of rainfall by hardwoods of eastern United States, *Wat. Resour. Res.* 1 : 193-206.

4. Jordan, C.F. 1968: Stemflow and nutrient transfer in a Traopical rain forest, *Oikos,* 31 : 257-263.

5. Nye, P.H. 1961: Organic matter and nutrient cycles under moist tropical forest. *Plant and Soil,* 13 : 333-346.

6. Raich, J.W. 1983: Throughfall and stemflow in mature and year old wet tropical forest 24(2) 234-243.

7. Sollins, P. and G. Drewry 1970: Electrical conductivity and flow rate of water through the forest canopy. In H.T. Odum and R.F. Pigeon (eds.) *A Tropical Rainforest.* U.S. Atomic Energy pigeon commission, Washington, D.C.

8. Yadav, A.K. and G.P. Mishra 1985: Chemistry of stemflow and throughfall waters for some tropical dry deciduous forest trees 1 pH, Specific conductivity, Nitrogen and phosphorus, *Journal of Tropical forestry.* Vol. 1(1) 51-60.

ENVIRONMENT AND EDIBLE OILS

A.S. Khalatkar

INTRODUCTION

Dynamics of population growth is always a matter of concern for planners of the resources and development in a country like India. Green revolution in 1968 has made India self-sufficient in food production, however, the growth was more quantity oriented. Now, the time has come to think about the quality, which is important not only from the point of view of the taste but for the overall health of the population. Edible oils are one of the major supplies of the much needed calories for the farm/factory labour and their output is directly dependent on the input in the form of food. In a developing country like India, it is essential to make available sufficient quantity of quality edible oil to maintain health and productivity. At present the major oilseeds in production are peanut, soybean, safflower, sesamum, cotton, etc. These oils are consumed by different sections of the population mainly for the sake of the flavour.

Edible oils are one of the major ecofriendly renewable energy resources. They are burnt as the fuel in human engines but are also being thought of as the future fuels to run mechanical engines. Hence, utilization of the edible oils has tremendous scope. However, this is possible only if the production levels keep on improving and are sufficient to satisfy the edible and other commercial usage.

Brassicas are one of the major oilseed groups in the family cruciferae and highly productive and input/output ratiowise beneficial to the farmers and can support the developing Indian economy. Four major brassicas are *Brassica napus, B. campestris, B. juncea* and *B. carinata.* Out of these four, two are widely grown *B. Juncea* and *B. campestris* by Indian farmers and are major contributors to Indian oilseed industry. *B. napus* internationally recognized as rapeseed is very popular quality oilseed with '00' characteristics. *B. carinata* is a relatively newer introduction in India and is productivity wise excellent but quality wise need further improvements.

Indian mustard, *B. juncea*, is characterized by > 50% erucic acid and > 135 μmol/g (deoiled cake) glucosinolates. Erucic acid is long chain unsaturated C : 22 fatty acid and is not considered nutritionally desirable. Similarly glucosinolates are the sulphus containing goitrogenic compounds and levels above 30 μm have been demonstrated to act as growth retardants in animals. Therefore both these compounds need to be low in oil for human consumption and in meal for animal feed. Indian mustard besides being high in these constituents is brown in colour and has > 30% protein and > 14% fibre. Oil extracted from brown seeds need bleaching and therefore it is necessary to have a yellow seeded mustard. This will eliminate the bleaching process and provide a genetic marker for distinguishing quality material. Further, yellow seed colour has been reported to reduce the fibre to 8% and increase protein > 38%. This adds to the nutritive value of the cake and is more desirable for animal feeds. Another important problem is the white rust susceptibility of Indian mustard. This disease results in a loss of about 25% in traditional areas and the crop needs heavy sprays of the fungicide polluting the environment. These sprays also add to the total input cost. Therefore it is essential to develop *B. juncea* varieties with genetic disease resistance.

MATERIAL AND METHODS

With these objectives my group in Department of Botany has undertaken induced mutation and conventional breeding programmes. In the first programme seeds of *B. juncea*, *B. napus* and *B. carinata* have been treated with sodium azide, ethyl methane sulfonate and gamma radiations. Treated seeds have been utilized for growing M1 generation. Individual M1 plants were harvested and half seeds were analyzed for fatty acid content by paper chromatography and GLC (Thies, 1971). Selected M1 seeds with low erucic acid were grown to raise M2 population. These were again harvested on M2 individual plant basis and screened for glucosinolate by tes-tape method (McGregor and Downey, 1975). Westar *(B. napus)* with < 30 μmol glucosinolate was used as the control. With such selection procedures it has been possible to generate the erucic acid and glucosinolate gene blocks in *B. juncea* and *B. carinata.*, *B. napus* CV Westar is already existing '00' variety with high productivity and oil, however, the crop takes approximately 165 days to mature and since the Indian winters, especially in Central India are not of longer duration the seed development gets affected, therefore the objective of our programme for *B. napus* is to develop an early flowering variety.

Quality breeding programme has recently received a boost due to the availability of zero erucic gene block ZYR 4 and Shiva (Tiwari *et al.*, 1989; Khalatkar *et al.*, 1991) and low glucosinolate (Love *et al.*, 1991). These lines were used in the reciprocal crosses. F1 seeds, were used to grow F2 generation. Individual F2 plants were selfed and harvested on individual plant basis. F2 half seeds were screened for fatty acid profile on GLC and paper chromatography. Selections were made for < 2% erucic acid and embryos were grown. Three week old seedlings were tested for white rust resistance in a temperature and humidity controlled growth chamber. This population was screened after two weeks of inoculation with the germinating spores of *Albugo candida* Race 2. Selections were made for the plants with strong white rust resistance based on the absence of pustule formation. This population with zero erucic acid and white rust resistance and yellow seed colour was transferred to the green house, selfed and harvested on individual plant basis. F3 seeds were sown in the field and multiplied. F3 plant progenies were again harvested on individual plant basis and screened for glucosinolate content. Selections were made only for plants which had introgressed zero crucic, yellow seed colour, white rust resistance and low glucosinolate characters.

RESULTS AND DISCUSSION

The data on saturated fat content in different oils are presented in Table 13.1. Canola oil obtained from rapeseed *(B. napus)* contains minimum saturated fat and is therefore nutritionally best balanced oil. By mixing low with high saturated fat oil different blends with medium saturated fat and flavours could be generated. This will benefit the consumer, oil industry as well as the farmer.

The induced mutation programme has made available hither to unavailable gene blocks and hybridization has made available interesting plants with '0' and '00' characteristics. The characteristics of these varieties are described in Table 13.2. These varieties can be commercialized immediately or they can be used as the basic breeding material for further improvements. The most important achievements of these investigations is to make available the quality material for edible oil industry and consumer health.

Table 13.1 : Saturated Fat Contents in Edible Oils Obtained from different sources.

Source	% Saturated Fat Content
Canola (Rape Seed)	6
Safflower	9
Sunflower	11
Corn	13
Soybean	15
Peanut	18
Cotton seed	27
Palm	57
Butterfat	66
Coconut	92

Table 13.2 : Important characteristics of newly bred *B. juncea*.

Varieties	Seed colour	White rust susceptibility	Oil content	Erucic acid %	Glucosinolate µ mole
Pusa bold	Brown	++++	36.6	52	124.4
Varuna	Brown	++++	38.0	53	179.9
Nihal	Brown	-	38.0	45	155.2
Shiva	Yellow	-	38.0	00	161.8
Heera	Yellow	-	38.0	00	26.4

Nihal is a variety in pusa bold background with white rust resistance, low eurucis acid and brown seed colour.

Shiva is characterized by yellow seed colour, white rust resistance, zero erucic acid and low glucosinolate. Heera is the variety with yellow seed colour, white rust resistance, zero erucis acid and low glucosinolate.

These materials hold a promise for the Indian quality breeding programme. Although these varieties are little late, and taller than the existing Indian varieties productivity wise they are on par.

Similarly, early flowering *B. napus* is also a '00' variety and by reducing its maturity period from 165 days to 146 days it can, now contribute to the Indian oil programme considerably. *B. carinata* material available till date is not of the same quality as *B. juncea* and *B. napus* and hence need more efforts to introgress quality characteristics.

Utilization of these quality materials will improve Indian edible oil and meal industry and bring good health to the consumer thereby contributing to the better environment and life expectancy. Enhanced oilseed production would also make a net energy deficit country energy sufficient and would provide energy for large scale development programmes.

ACKNOWLEDGEMENTS

The author is thankful to CIDA/NSERC, Canada, BRNS, Department of Atomic Energy and National Dairy Development Board for financial assistance.

REFERENCES

1. Khalatkar, A.S., Rakow, G. and Downey, R.K. 1991. Development of white rust resistant, yellow seeded, zero erucic and Indian mustard *(Brassica juncea)* In : Proc. Eighth Internat. Rapeseed Cong. Saskatoon, Canada, 5 : 1475-1477.

2. Love, H.K., Rakow, G., Raney, J.O. and Downey, R.K. 1990. Development of low glucosinolate mustard. Can. J. Plant Sci., *70* : 419-424.

3. McGregor, D.I. and Downey, R.K. 1975. A rapid and simple assay for identifying low glucosinolate rapeseed. Can. J. Plant Sci. *55* : 191-196.

4. Thies, W., 1971. Rapid and simple analysis of fatty acid composition of individual rape cotyledons. I. Gas and Paper Chromatography. Z. Flazenzuchtg. *65*(3) : 181-202.

5. Tiwari, A.S., Petrie, G.A. and Downey, R.K. 1988. Inheritance of resistance to *Albugo candida* race 2 in mustard *(Brassica juncea* (L) (Czern), Can. J. Plant. Sci. *68* : 297-300.

DETERIORATION OF POLLEN FERTILITY IN RESPONSE TO AIR POLLUTION - A REVIEW

S.V.S. Chauhan and Jolly Singh

ABSTRACT

The work on deterioration of pollen fertility in flowering plants in response to air pollution has been reviewed. It covers the effect of ozone, fluoride and sulphur dioxide on pollen fertility in 43 species, 38 genera and 21 families of angiosperms and 5 species, 2 genera and 1 family of gymnosperms.

INTRODUCTION

The pollen grain is the discrete and mobile stage of the highly reduced male gametophyte of higher plants and it is shed in a highly desiccated condition (Frankel and Galun, 1977). Pollen transmits the male genetic material in sexual reproduction of all higher plants and its important role in the formation of seed was demonstrated by Camerarious as early as 1694 by his elegant experiment (Shivanna and Johri, 1985). Pollen is also well suited as a research tool for studying many patterns of plant and animal metabolism and an increased knowledge of pollen may help plant breeders accelerate efforts to improve the world's food and fibre supply. The development of haploid pollen takes place in a diploid environment and the tapetum, the inner most anther wall layer provides both nutrition for developing pollen grains and helps in anther dehiscence (Chauhan, 1977). On anther dehiscence pollen grains are shed in the external environment and the pollen has to survive before it lands the receptive surface of stigma. During this transference, the pollen is entirely at the mercy of a large number of abiotic (air flow, temperature, RH, rainfall etc.) and biotic factors (insects, flowers, nectar of the flower) and use of various gametocides (see Chauhan and Kinoshita, 1982), pesticides, insecticides, fungicides and pollutants as well (Shivanna and Johri, 1985) have been instrumental in reducing pollen fertility. Pollen fertility of a crop is of significant importance to plant breeders, because reduction in fertility serves as a limiting factor in hybridization programmes and it also affects the yield of crops. In the present article an attempt has been made to review the literature on the deterioration of pollen fertility in plants due to air pollution through 1994.

EFFECT OF AIR POLLUTION ON POLLEN FERTILITY

1. *Effect of fluoride*

Fluorides in the atmosphere act as pollutants and severely affect fruit and seed formation. Schulzbach and Pack (1972) have reported that tomato plants grown in the presence of hydrogen fluoride (HF) produced smaller fruits which were partially or totally seedless. They have also reported that in tomato and cucumber, sodium fluoride (NaF) inhibited *in vitro* pollen germination.

Facteau *et al.* (1973) and Facteau and Rowe (1981) have reported that in *Prunus avium* increased fluoride fumigation levels resulted in a decrease in per cent pollen germination and pollen tube growth.

2. *Effect of Ozone*

Feder (1981) has successfully used *in vitro* culture of tobacco, petunia and tomato pollen for bioassay of ozone. The pollen tube growth was found to be highly sensitive to low concentrations of ozones.

3. *Effect of Sulphur dioxide*

Dopp (1934) has found that moist pine pollen grains exposed to 10 ppm SO_2 for 45 minutes loose their fertility as indicated by reduction in *in vitro* pollen germination. He has also observed that the exposure of pollen grains of *Digitalis purpurea* to SO_2 was more effective on *in vitro* than *in vivo* pollen germination. Moist pollen grains of *Pinus montana, P. deltoides, P. resinosa, P. nigra, Picea pungens* when exposed to 10 ppm of SO_2 for 45 minutes, their pollen tube elongation was considerably reduced.

On the other hand, Antipov (1970) has reported that pollen germination in woody plants was stimulated by exposure with SO_2.

Karnosky and Staris (1974) have reported that germination of pollen grains of *Populus deltoides, Pinus nigra, P. resinoda* was reduced when exposed for 4 hours in moist condition to 0.75 ppm SO_2. Similarly, pollen grains of *Pinus resinosa, P. nigra* and *Picea pungens* exposed to 1.4 ppm SO_2 exhibited a marked reduction in pollen tube length. They have also observed frequent bursting of pollen tubes.

Ma and Khan (1976) have reported that the pollen grains of *Tradescantia* when exposed to high concentrations of SO_2 (10, 50, 100, 250, 750, 1000 and 2000 ppm) exhibit reduction in pollen tube growth which was directly proportional to increase in the concentration of SO_2. They have also observed inhibition of mitotic activity of the generative nucleus in the developing pollen tube culture of *Tradescantia axillaris* by low levels of SO_2 (1310 µg/m^3) when exposed for brief period of 2-4 hours.

Masaru *et al.* (1976) have observed that on exposing pollen grains of *Lilium longiflorum* to various concentrations of SO_2, CO_2, HCHO and CH_3=CH-CHO singly and in various combinations for 1, 2 and 5 hours markedly inhibited pollen tube elongation. Combination of 0.75 ppm SO_2 and 240 ppm formaldehyde was most effective.

Linzon (1978) has found that fertilization and pollen tube growth was retarded by the influence of air pollutant i.e. sulphur dioxide.

In the opinion of Varshney and Varshney (1981) the pollen grains are highly sensitive to SO_2 and loss in their germination percentage can be used for detecting SO_2 pollution. Pollen germination in *Petunia alba* and *Tradescantia axillaris* exposed to 131 µg/m^3 SO_2 under moist conditions was 52 and 49%, respectively, while, in *Cicer arietinum* and *Nesteratium indicum,* it was only 4% and 8% respectively. At 262 µg/m^3 SO_2, the germination was uniformly low in all the species and at 1310 µg/m^3 SO_2, pollen grains fail to germinate. Over 95% reduction in pollen tube growth was observed in *C. arietinum* when exposed to 131 µg SO_2/m^3 for 3 hours and 97% reduction has been reported in case of *N. indicum, P. alba* and *T. axillaris* when exposed to 262 µg/SO_2/m^3 for 3 hours.

Dubey (1983) has reported that combined stress of herbicides and sulphur dioxide pollutants induce server reduction in pollen germination and pollen tube growth. The experiment with herbicides and SO_2 individually indicated inhibition of pollen germination and tube growth in field and laboratory conditions. The stress of herbicides and SO_2 affected the pollen severely as only 26 and 31% respectively could germinate in

Solanum melongena. Pollen tube length was 1.69, 2.18 and 1.0 mm in plants treated with herbicides (pendimethalin, Asulox) and SO_2 (600 μg m^{-3}) respectively. However, plants treated with herbicides and sulphur dioxide showed 0.28, 0.29 mm long pollen tubes respectively.

Singh and Chauhan (1988) have made a comparative survey of pollen fertility of some plants *(Bougainvillea glabra, Brassica campestris, Cajanus cajan, Cassia fistula, Convovulus pluriculis, Croton bonpladianum, Ipomea fistulosa, Parkinsonia aculeata, Prosopis jubiflora, Salvadora olesides, Salvadora persica* and *Solanum xanthocarpum)* growing around Mathura refinery and Vindravan forest. According to them, the plants growing around the polluted area of the refinery exhibited significantly poor pollen fertility as compared to those growing in non-polluted area of Vindravan forests. Vardharajan (1977) and Calloway *et al.* (1978) have reported that major gas pollutants produced by the Mathura Refinery unit are oxides of sulphur, hydrogen sulphide, oxides of nitrogen, hydrocarbons, ammonia and carbon monoxide.

Jain and Chauhan (1989) have explained that plants, are bioindicators of air pollution and loose their fertility in stress conditions. The pollen fertility of *Raphanus sativus, Cosmos sulphureus, Lablab purpureus, Jatropha gossypifolia* fumigated with 1 ppm SO_2 for 2 hours, in glass desiccators was adversely affected. The marked reduction in fertility of pollen exposed to SO_2 was 12.35, 10.71 and 10.18% recorded in *Brassica campestris, Clerodendron splendensand Catharanthus roseus* plants respectively.

Bharadwaj and Chauhan (1989) have found that *Nerium odorum* plants growing in polluted area which carried a heavy traffic of automobiles, exhibited poor pollen fertility, tube elongation and increase in the time of emergence of pollen tube.

Jain *et al.* (1989) have made a comparative study on the rate of floral abscission and pollen fertility in some ornamental plants growing in an area with a large number of automobile work shops. These results were compared with some species of plants growing in the College garden. According to them, the plants of *Nyctanthus arbortristis, Tecoma stans, Quisqualis indica, Cassia sophera, Jasminum multiflorum, Bougainvillaea spectabilis* growing around the automobile work shops exhibited significantly high rate of floral abscission and poor pollen fertility as compared to those growing in the garden.

Singh *et al.* (1990) have reported that plants of *Cajanus cajan, Brassica nigra* and *Acacia nilotica* growing around Mathura Refinery showed a marked reduction in their pollen fertility.

Chauhan and Jain (1990) have observed that *Brassica campestris* plants growing at different sites of National High Way No. 2 showed considerable reduction in the fertility of pollen in the area near the Refinery, barari, Refinery, Flyover and Refinery Colony (33.54, 69.67, 81.26 and 61.24% respectively), as compared to that of plants growing at non-polluted area (Botanical garden, R.B.S. College, Agra 84.67%).

Jain and Chauhan (1990) have also reported that fumigated plants of *Capsicum annuum* exhibited a marked reduction in pollen fertility which increased with the increase in the concentration of SO_2 and the time of emergence of pollen tube was also enhanced.

Jain and Chauhan (1990) have observed that the pollen grains of *Coccinia indica, Impatiens balsamina, Lablab purpureus* and *Parkinsonia aculeata,* when fumigated with 0.5 ppm SO_2 in a glass desiccator for 10 minutes lost their fertility as pollen tube formation and tube growth was considerably reduced.

Chauhan and Jain (1994a) have observed the effect of 0.25, 0.50 and 1 ppm SO_2 on pollen fertility in *Capsicum annuum* L. var. Pusa jwala. The plants fumigated in one cubic meter open top polythene chambers with 0.25, 0.50 and 1 ppm SO_2 for 1 ppm SO_2 for 4 hours a day at an interval of 24 hours for 25, 50, 70 and 100 days at seedling, vegetative and bud initiation stage exhibited a marked reduction in their pollen fertility. The reduction in pollen fertility increased with the increase in the concentration and period of fumigation. Maximum reduction was shown by plants fumigated with 1 ppm SO_2 for 50 days at bud-initiation stage.

Chauhan and Jain (1994b) have studied the effect of SO_2 fumigation on *in vitro* pollen germination in *Capsicum annuum*. The plants fumigated with different concentrations of SO_2 for various periods, exhibited marked reduction in *in vitro* pollen germination percentage and pollen tube elongation. Pollen of plants fumigated with 1 ppm SO_2 for 100 days at seedling stage showed only 28.4% *in vitro* pollen germination while the shortest pollen tubes (1345 um) developed from pollen of plants exposed with 1 ppm SO_2 for 75 days at vegetative stage.

Chauhan and Jain (In Press) have observed that the plants of *Capsicum annuum* fumigated with SO_2 exhibited a marked reduction *in vivo* pollen germination. Fumigation with 1 ppm SO_2 for 100 days starting from seedling stage showed only 12.8% germination on stigmatic surface and the pollen tubes were only 105 um long as compared to unfumigated plants showing 59% germination with 270 um long tubes.

DISCUSSION AND CONCLUSION

This review pertains to the effect of ozone, fluoride and sulphur dioxide on pollen fertility deterioration in a wide variety of tax of both gymnosperms and angiosperms. It is quite evident from the studies made by a large number of workers that various gases, SO_2, O_3 and fluoride exhausted from industries and automobiles in the atmospheric cause considerable reduction in the fertility of pollen in different plants including important crops. Since the pollen grains are highly sensitive as they quickly loose their fertility in the atmosphere containing various gases, particularly sulphur dioxide. According to Das and Runneckles (1974), Jagiello *et al.* (1975) and Petering and Shih (1975) the higher toxicity of sulphur dioxide to moist pollen grains appears to be due to conversion of SO_2 into SO_3, HSO_3 ions or in the opinion of Ma *et al.* (1973) and Ma and Khan (1976) it may be due to dissoluted but unreacted form of sulphur dioxide i.e. SO_2 H_2O. The reason for the higher sensitivity of pollen tube growth is not yet clear. However, from the available information it can be speculated that if may be due to the effect of sulphur dioxide on the generative nucleus. Sulphur dioxide appears to affect the DNA metabolism in the generative nucleus (Ma *et al.* 1973). This is substantiated by the observations of Khan and Ma (1974) on the effect of 3 mm hydroxyurea an inhibitor of DNA synthesis in pollen grains. Pollen tube growth was reduced by 5 mm solution of hydroxyurea and the response was comparable with inhibitory effect of 131 $\mu g/m^3$ of sulphur dioxide. Thus, the pollen grains of some sensitive plant species may be effectively used as bioindicators of air pollution and can also be used to detect SO_2 pollution. Based on the findings of others presented in support, if can be concluded that the reduction in the pollen fertility may be due to abnormal changes in the DNA metabolism in the generative nucleas. The pollen grains, the male gametophyte is less protected and is nacked and loose their fertility quickly under all kinds of stress (environmental, Chemical or even pathological).

REFERENCES

1. Antipov, V.G. 1970. The effect of SO_2 on the reproductive organs of woody plants *Ohrana priodyna* vrale Sterdlovak No. 7 : 31-35. Forest Abstr. 32: 4 : 752.

2. Bharadwaj, G. and Chauhan, S.V.S. 1989. Effect of air pollutants from automobiles on pollen fertility in *Nerium odorum* L. Acta Ecol. 11 : 1 : 34-37.

3. Calloway, J.A., Schwartz, Jr. A.K. and Thompson, R.G. 1978. In the cost of Energy and a clean Environment Thompson, R.G. Calloway, J.A. Nawalanic, L.A. Eds. Gulf Publishing Co. Houston, U.S.A.

4. Chauhan, S.V.S. 1977. Dual role of tapetum. Curr. Sci. 46 : 489-490.

5. Chauhan, S.V.S. and Jain, P.K. 1990. Pollen fertility of *Brassica campestris* growing at different sites on national high way No. 2. Acta Ecol. 12 : 2, 77-82.

6. Chauhan, S.V.S. and Jain, P.K. 1994a. Effect of SO_2 fumigation on pollen fertility in *Capsicum annuum* L. Acta Ecol. 16, 30-33.

7. Chauhan, S.V.S. and Jain, P.K. 1994b. *In vitro* pollen germination in *Capsicum annuum* plants fumigated with SO_2. J. Environ. Pollut. 1(3-4).

8. Chauhan, S.V.S. and Jain, P.K. (in Press). *In vivo* pollen germination in chillies *(Capsicum annuum)* fumigated with sulphur dioxide. Japan J. Palynol.

9. Chauhan, S.V.S. and Kinoshita, T. 1982. Chemically induced male sterility in angiosperms Seiken Ziho 30 : 54-75.

10. Das G. and Runneckles, V.C. 1974. Effect of bisulfite on metabolic development in synchronous *Chlorella pyrenoidosa* Environ. Res. 7 : 353-362.

11. Dopp, W. 1934. Uber die Wirkung der Schwe filgen sure auf Blutenorgne. Ber. deut. bot. Ges. 49 : 173-221.

12. Dubey, P.S. 1983. Additive toxicity of SO_2 and herbicides, a new air pollution problem. In proceeding of the symposium on Air Pollution Control New Delhi Indian Association for Air Pollution Control 1 : 115-118.

13. Facteau, T.J. and Rowe, K.E. 1981. Response of sweet cherry and apricot pollen tube growth to high levels of sulphur dioxide. J. Am. Soc. Hortic. Sci. 106 : 77-79.

14. Facteau, T.J.; Wang, S.Y. and Rowe, K.E. 1973. The effect of hydrogen fluorise on pollen germination and pollen tube growth in *Prunus avium* L. Cv. 'Royal Ann'. J. Am. Soc. Hortic. Sci. 98 : 234-236.

15. Feder, W.A. 1981. Bioassaying for ozone with pollen system. Environ. Hlth. Persp. 37 : 117-123.

16. Frankel, R. and Galun, E. 1977. Pollination mechanisms, reproduction and breeding. Springer-Verlag, Berlin, Heidelberg, New York.

17. Jagiello, G.M., Lin, J.S. and Dacayen, M.P. 1975. SO_2 and its metabotite: Effect on mammalian egg chromosomes. Environ. Res. 9 : 84-93.

18. Jain, P.K. and Chauhan, S.V.S. 1989. Impact of SO_2 fumigation on pollen grains of some plants. Incompatibility Newsletters. 21 : 31-33.

19. Jain, P.K. and Chauhan, S.V.S. 1990. Effect of SO_2 fumigation on flowering pollen fertility and fruiting in *Capsicum annuum* L. Incompatibility Newsletter. 22 : 15-17.

20. Jain, P.K. and Chauhan, S.V.S. and Sumitra Kumari, 1989. "Impact of air pollution on floral abscission and pollen fertility in some ornamental plants". Acta. Ecol. 11(1) : 31-33.

21. Khan, S.H. and Ma, T.H. 1974. Hydroxyurea enhanced chromated abberrations in *Tradescantis* pollen tubes and seasonal variation of aberration rates. Mutal. Res. 25 : 33-38.

22. Karnosky, D.F. and Staris, G.R. 1974. The effect of SO_2 on *in vitro* forest tree pollen germination and tube. elongation. J. Envrn. Qual. 3 : 406-409.

23. Linzon, S.N. 1978. Effect of air borne sulphur pollutants. In : sulphur in environment. Ebs. by J.O. Nriagy Part II. Wiely, New York, 109-162.

24. Ma, T.H. and Khan, S.H. 1976. Pollen mitosis and pollen tube growth inhibition by SO_2 in cultured pollen tubes of *Tradescanlia* pollen tubes and seasonal variation of the aberration rates. Mutal. Res. 21 : 93-100.

25. Ma, T.K., Isbandi, D.; Khan, S.H. and Tseng, Y.S. 1973. Low level of SO_2 enhanced chromated abberrations in *Tradescantis* pollen tubes and seasonal variation of the abberration rates. Mutal. Res. 21 : 93-100.

26. Masaru, N.; Syozo, F. and Salwro, K. 1976. Effect of exposure to various injurious gases on germination of lily pollen, Environ. Pollut. 11 : 181-187.

27. Petering, D.H. and Shih, N.T. 1975. Biochemistry of bisulphite sulphur dioxide. Environ. Res. 9 : 55-65.

28. Schulzbach, C.W. and Pack, M.R. 1972. Effects of fluoride on pollen germination, pollen tube growth, and fruit development in tomato and chumber. Phytopathology 62 : 1247-1253.

29. Shivanna, K.R. and Johri, B.M. 1985. The Angiosperm Pollen structure and Function. Wiley Eastern Limited, New Delhi.

30. Singh, D. and Chauhan, S.V.S. 1988. Effect of air pollution on pollen fertility in some plants. Acta. Ecol. 10(2) : 68-71.

31. Singh, D., Faroqui, F. and Dhakre, J.S. 1990. Effect of air pollutants on chlorophyll content and pollen fertility of some plants around Mathura Refinery. Acta. Ecol. 12 : 1-34-39.

32. Vardharajan, S. 1977. Report of expert committee on environmental impact of Mathura Refinery. pp. 12-15.

33. Varshney, S.R.K. and Varshney, C.K. 1981. Effect of sulphur dioxide on pea seedlings. Indian J. Air Pollut. Cont. Vol. 2. a : 47-49.

SOME AEROMYCOLOGICAL POLLUTANTS OF MEAT MARKET

Akhilesh Ayachi, R.K. Shrivastava, Bhawna Thakur and O.P. Chourasia

ABSTRACT

The present paper deals with the study of aerospora of fungal origin over a meat market. The sampling has been done by settle plate method and using Rotorod sampler during Oct. 1993 to March 1994. A total of thirty two fungi were isolated. The dominant being *Aspergillus, Cladosporium, Curvularia, Fusarium, Alternaria, Helminthosporium* and *Rhizopus* etc.

Five fungal forms known to be highly pathogenic and causing mycoses in human beings were also isolated.

INTRODUCTION

Biotic environmental pollutants comprise chiefly of fungal spores. The dead and decaying organic materials, animal products and plants contribute the fungal propagules into the atmosphere (Ayachi *et al.*, 1991, 1992).

Many human diseases are attributed to air borne fungal spores. Chaubal *et al.*, (1985).

Rajak *et al.*, (1989) reported the presence of forty two fungal forms from the soil of gelatin factory campus, Jabalpur. Majority of the fungi belonging to different genera were found keratinophlic in nature. In factory campus raw bones are generally placed scattered in open fields. The degradation of flesh attached to bones was reported by variety of micro-organisms. A number of these micro-organisms are harmful to health as they initiate many skin, respiratory and allergenic diseases.

So far no attempt has been made to ascertain the air mycoflora of meat markets. To fill the lacuna in our knowledge of aeromycology a meat market was selected for present study.

MATERIALS AND METHODS

The aeromycoflora of Bhantaliya meat market, Jabalpur was recorded by settle plate method. Petriplates of potato dextrose agar media were exposed to the atmosphere of meat market twice a week, for 5 minutes at 75 cm. above ground level for qualitative estimation of aeromycoflora. The plates were incubated for 5-7 days at 28°C, after which the developing colonies were transferred to agar slants for characterization and identification.

Rotorod sampler of Parkins (1957) modified by Harrington (1959) was employed to study the aeromycoflora in meat market. The sampler was operated twice a day (morning and evening) with an interval of four days. The method of sampling and scanning was followed as described by Tilak and Srinivasulu (1967).

RESULTS

A total of 32 fungi were isolated (Table 15.1). The dominant fungal spores in the air of meat market were represented by *Aspergillus, Cladosporium, Curvularia, Fusarium, Alternaria, Helminthosporium,* and *Rhizopus* (Fig. 15.1).

Table 15.1 : Aeromycoflora of a meat market by using Petriplate method Sampling Period 1993-94

		Oct.	*Nov.*	*Dec.*	*Jan.*	*Feb.*	*Mar.*
	Achaetomium sp.	-	-	-	++	++	+
*	*Acremonium* sp.	-	-	-	+	+	++
	Aurobasidium pullulans	-	-	+	+	-	-
	Alternaria alternata	+	+	+	+++	++	+
	A. Tenuis	+	+	-	+	+	-
	Aspergillus flavus	-	-	-	-	+	+
*	*A. fumigatus*	-	-	-	-	+	-
	A. niger	-	+	+	+	+	-

	Oct.	*Nov.*	*Dec.*	*Jan.*	*Feb.*	*Mar.*
A. terreus var. *aureus*	+	+	-	-	+	+
A. Sydowi	+	+	+	+	+	+
* *Candida albicans*	-	-	-	+	+	-
Cladosporium cladosporoides	-	-	-	+	+	+
C. herbarum	+	+	+	+	+	+
Chaetomium sp.	-	-	+	+	+	-
Coleophoma sp.	-	-	-	-	+	-
Conanephora cucurbitarum	+	-	-	-	-	-
Curvularia lunata var. aeria	-	-	-	-	+	+
C. pallescens	+	-	-	+	-	+
Dicotomyces sp.	-	-	-	-	-	+
Fusarium nivale	+	-	-	-	-	-
F. rosium	-	+	-	-	-	+
Helminthosporium sp.	-	+	+	-	+	-
Humicola sp.	+	+	+	-	-	-
* *Microsporum* sp.	-	-	-	+	+	-
Melanospora sp.	-	-	-	-	++	+
Myrothecium roridum	+	-	-	-	-	-
Phoma sp.	-	-	-	+	-	+
* *Rhizopus* sp.	-	-	-	-	+	+
Scytilidium sp.	-	-	-		+	+
* *Sporothrix* sp.	-	-	-	+	+	-
White sterile colonies	+	+	+	+	+	+
Black sterile colonies	+	-	+	-	+	-

* Pathogenic forms

Symptoms of fungal diseases (mycoses) were usually seen in people inhabiting the areas near this meat market.

DISCUSSION

The presence of keratinophilic fungi such as *Candida albicans, Microsporum* sp. *Aspergillus fumigatus, Sporothrix* sp. and *Acremonium* sp. in the air of meat market area are perhaps due to the presence of hairs,

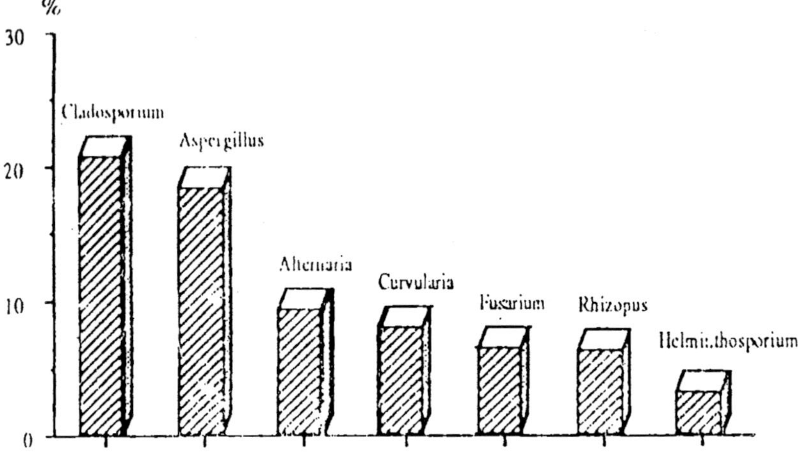
%
30
20
10
0
Cladosporium
Aspergillus
Alternaria
Curvularia
Fusarium
Rhizopus
Helminthosporium

Fig. 15.1

bones and discarded waste of meat in the meat market, which act as bait in soil and this increases the inoculum potential of these isolated keratinophilic fungi and gets directly exposed to the air currents. Gillman (1959) and Mukerjee (1966) suggested that 90% of the fungal spores present in the air are those which are soil inhabitants.

In the present study the presence of keratinophilic pathogens forms in the air of meat market are normally present in the soil of meat market area was in conformity by other workers that, in normal course, the spores are blown/flown from soil (Ingold, 1971).

In the present survey, the symptoms of fungal disease (mycoses) were usually seen on people inhabiting in nearby area of this meat market. Grin and Ozcoric (1963) observed similarities between saprophytic keratinophilic fungi and obligate skin dermatophytes which have led to speculation about their pathogenic nature. Among these fungal groups producing diseases, only keratinophilic fungi have shown a parasitic mode of survival.

Chattaway *et al.*, (1963), Yu *et al.*, (1972), Takaiuchi *et al.*, (1982) suggested that keratinophilic fungi are toxic and their toxicity causes allergenic reactions.

Tilak (1981) and Ayachi (1991) stated that, in aeromycological studies, source for release of viable spores from soil or other sources and their deposition on suitable surfaces, and germination give an idea how pathogenic fungi spread and cause diseases. In the present study also similar situation was observed.

ACKNOWLEDGEMENTS

Grateful thanks are due to Dr. Shiv Shankar Ayachi, Professor and Head, Deptt. of Botany and Environmental Science, Govt. Autonomous Science College, Jabalpur, for their valuable suggestions. Thanks are also due to Principal, Govt. Autonomous Science College, Jabalpur for providing laboratory facilities.

REFERENCES

1. Ayachi, A. and Rai, B. (1991). Aeromycological studies on grade yard in relation to Hindu Traditions. 6th National conference on Aerobiology.

AIRBORNE POLLEN GRAINS AND DOMINANT FUNGAL SPORE TYPES OF THE WINTER

R.P. Mishra

ABSTRACT

This chapter presents biweekly fluctuation in the occurrence of pollen grains (20 types) and certain dominant fungal spores (12 types) in the atmosphere of Jabalpur. Air sampling was carried out by modified Durham's air sampler during December, 1983, to February, 1984. Dry period with low R/H favoured the occurrence of pollen grains whereas wet period with high R/H favoured the incidence of fungal spores. No significant correlation was found with atmospheric temperature.

INTRODUCTION

Airborne pollen grains and fungal spores are chief causative agent of respiratory allergy. Several airborne fungi are also pathogenic to human beings, animals and plants. An aerobiological information regarding changes in the concentration and composition of allergens is needed especially for medical purposes, on account of forecasting and treatment of allergic manifestations. Knowledge about the prevalence of phytopathogenic inoculum in the air is also useful for establishing forecasting system for plant diseases. The aerobiological data having such a multipurpose utility have been collected throughout the world including India. In Jabalpur, such studies were pioneered by Agarwal *et al.* (1983). Later on, an array of workers made considerable attention towards studies on outdoor (Khare, 1984; Mishra, 1985, 1987; Verma & Khare 1987a) and indoor (Verma & Khare, 1987 b; Verma & Pandey, 1992; Verma *et al,* 1992) air spora of Jabalpur.

Jabalpur city is situated at 23°-10'N and 79°-57'E at a height of 402 m MSL. The city is decorated with hilly tract covered by luxuriant vegetation. It has a tropical climate with four different seasons viz. monsoon, post-monsoon, winter and summer. The winter season often faces a phenomenon of "temperature inversion" which is characterised by the higher concentration of airborne bioparticles in the lower atmospheric layers. This situation is responsible for the prevalence of respiratory allergy especially chronic asthma. The present study was therefore, undertaken to picturise the biweekly occurrence of aeroallergens/pathogens in relation to rainfall, atmospheric temperature and relative humidity.

MATERIAL AND METHODS

Modified Durham's air samplers were used for the trapping of air spora. The original design of Durham's sampler was modified in such a way that one of the exposed slides faces wind tunnels at 30° angle. Air sampling was carried out during December, 1983 to January, 1984 by installing three samplers i.e. two at Art Block of R.D. University campus at 4.58 m (15) feet) and 13.73 m (45 feet) height and one at Bai ka Bagicha (13.73 m height). These sites are 5 km away from each other. The vaseline coated slides (two in each sampler) were exposed at 10 a.m. continuously for two/three days. The slide having maximum deposition (one from each sampler) was scanned under high magnification (450 X). The pollen/ fungal spores trapped in an area of 18×18 mm^2 in each slide were identified and counted. Identification was made with the help of available literature and standard reference slides. The meteorological data were collected from the Department of Meteorology, Adhartal, Jabalpur.

RESULTS AND DISCUSSION

Survey revealed 20 types of pollen belonging to 16 angiospermic families and 12 dominant spore types out of 80 identified fungal spores (Plate 16.1 & 16.2). The majority of the trapped pollen/fungal spores is of known allergenic potential (Table 16.1). The seasonal census of trapped pollen grains and fungal spores revealed a ratio of 1 : 87. Maximum contribution (22%) to the air pollen flora was made by *Eucalyptus* (Myrtaceae) and minimum (0.14%) by *Blepharis* (Acanthaceae). Tree pollen were encountered more followed by herbs and shrubs. Frequency of herb pollen decreased gradually from December to January. No such decrease was found in the pollen of trees and shrubs. Fig. 16.1 representing total trapping of three locations reveals that maximum number of pollen grains occurred during December (rainfree month), followed by February (with 5 rainy days) and January (with 8 rainy days). Thus, rain as well as

Plate 16.1 : Camera lucida drawing of the pollen grains trapped from the atmosphere

Fig. 1 = *Ailanthus excelsa*

2 = *Albizia lebbeck*

3 = *Amaranthus*

4 = *Arthemisia nilagirica*

5 = *Blepharis*

6 = *Bougainvillea*

7 = *Cassia siamea*

8 = *Cassia occidentalis*

9 = *Ceiba pentandra*

10 = Grass pollan

11 = *Jpomoea fistulosa*

12 = *Jasminum*

13 = *Eucalyptus*

14 = *Lantana*

15 = *Millingtonia hortensis*

16 = *Moringa oleifera*

17 = *Parthenium hysterophorus*

18 = *Peltophorum pterocarpum*

19 = *Plumeria*

20 = *Tridax procumbens*

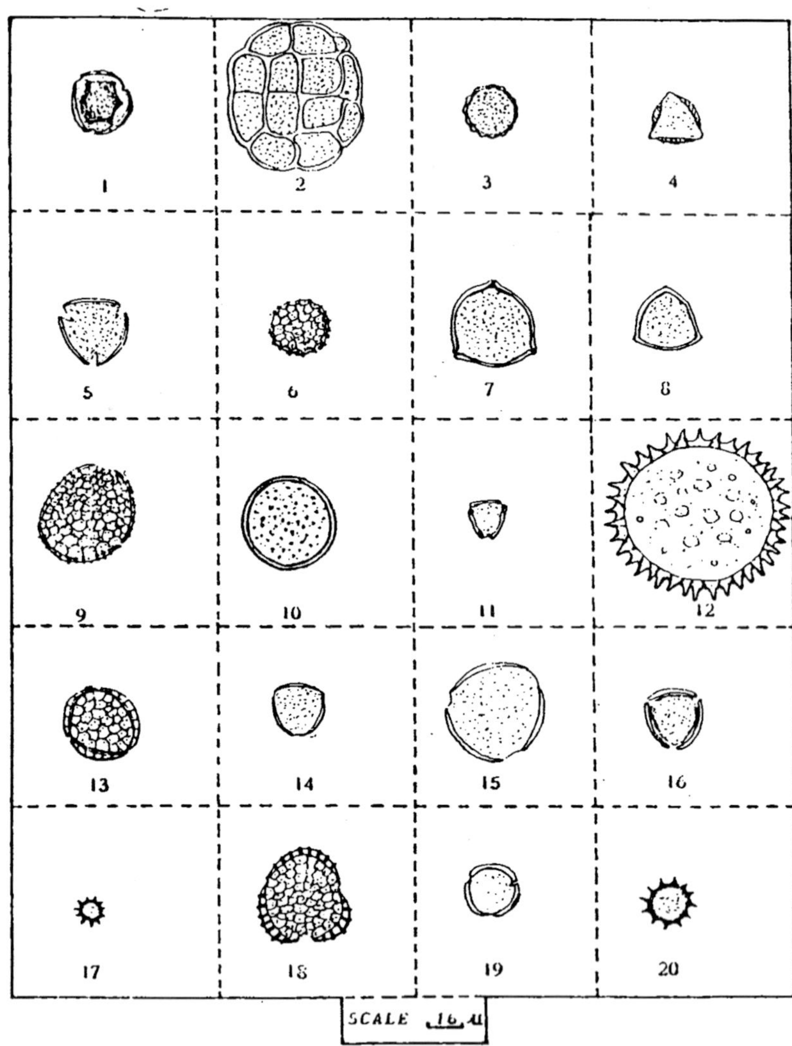

Plate 16.1

Plate 16.2. Camera lucida drawings of certain dominant fungal spores trapped from the atmosphere

Fig. 1 = Aspergilli-Penicilli

2 = *Cladosporium*

3 = Rust spores

4 = *Botrytis*

5 = *Alternaria*

6 = Smut spores

7 = *Periconia*

8 = Mucoraceae spores

9 = Basidiospores

10 = *Bispora*

11 = *Curvularia*

12 = *Nigrospora*

13 = *Helminthosporium*

14 = *Pleospora*

15 = *Epicoccum*

16 = *Pithomyces*

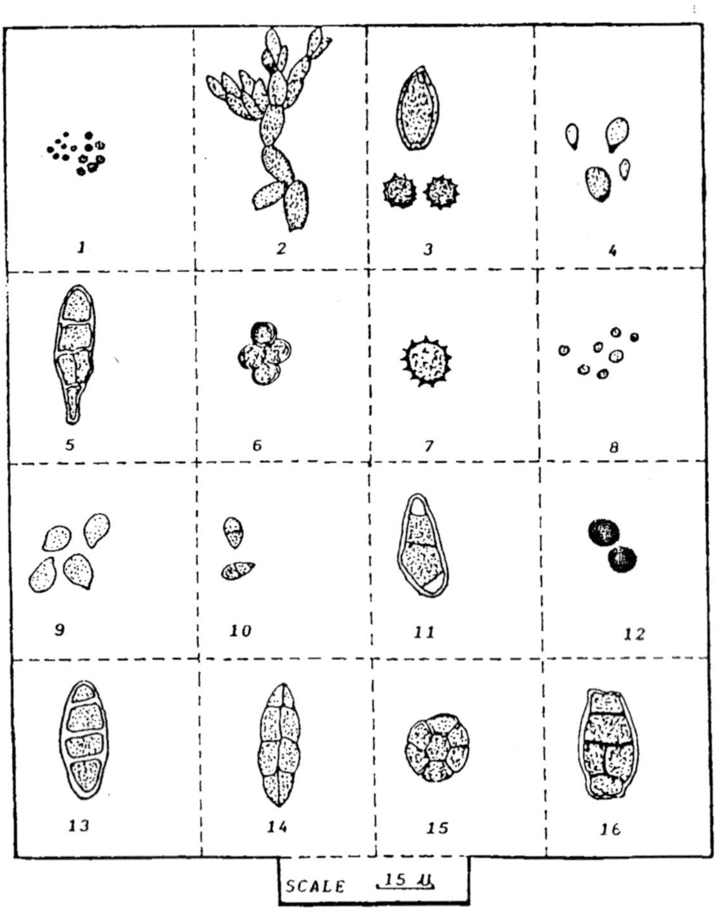

Plate 16.2

Table 16.1 : Total monthly count of pollen grains and fungal spores trapped at three different locations

Pollen/spore	Number				% Out of group Total
	Dec.	Jan.	Feb.	Total	
Pollen Type					
* *Eucalyptus*	279	34	4	317	22.00
* Grass pollen	126	55	56	237	16.41
* *Ailanthus excelsa*	-	4	217	221	15.30
* *Lantana*	53	51	51	155	10.73
* *Cassia siamea*	39	35	71	146	10.11
Peltophorum pterocarpum	72	2	-	74	5.12
Ceiba pentandra	6	-	31	37	2.56
Albizia lebbeck	-	-	31	31	2.15
* *Tridax procumbens*	25	4	-	29	2.01
Moringa oleifera	20	-	-	20	1.39
* *Parthenium hysterophorus*	18	-	-	18	1.25
Cassia occidentalis	12	-	-	12	0.83
Artemisis nilagirica	12	-	-	12	0.83
Millingtonia hortensis	10	-	-	10	0.69
* *Amaranthus*	5	4	-	9	0.62
Jpomoea fistulosa	-	8	-	8	0.55
Jasminum	-	5	-	5	0.35
Bougainvillea	-	3	-	3	0.21
Plumeria	-	-	3	3	0.21
Blepharis	-	2	-	2	0.14
Unidentified	21	68	6	95	6.58
Total Pollen	693	275	476	1444	100.00

1	*2*	*3*	*4*	*5*	*6*
Fungal Spore Type :					
* Aspergilli-Penicilli	486	53393	25813	79792	63.08
* *Cladosporium*	2315	9805	4345	16465	13.04
* Rust spores	55	10694	339	11188	8.86
* Basidiospores	444	3444	1055	4943	4.21
* *Alternaria*	521	368	1193	2082	1.65
* Smut spores	175	984	651	1810	1.43
* *Periconia*	209	619	923	1751	1.39
* Mucoraceous spores	96	1461	55	1612	1.04
Botrytis	-	252	1025	1277	1.01
* *Curvularia*	278	254	128	660	0.52
* *Nigrospora*	170	148	211	529	0.42
Helminthosporium	113	104	168	385	0.30
Other type	678	2054	1109	3841	3.05
Total type	5540	83680	37014	126234	100.00

- = Absent * = well known allergens.

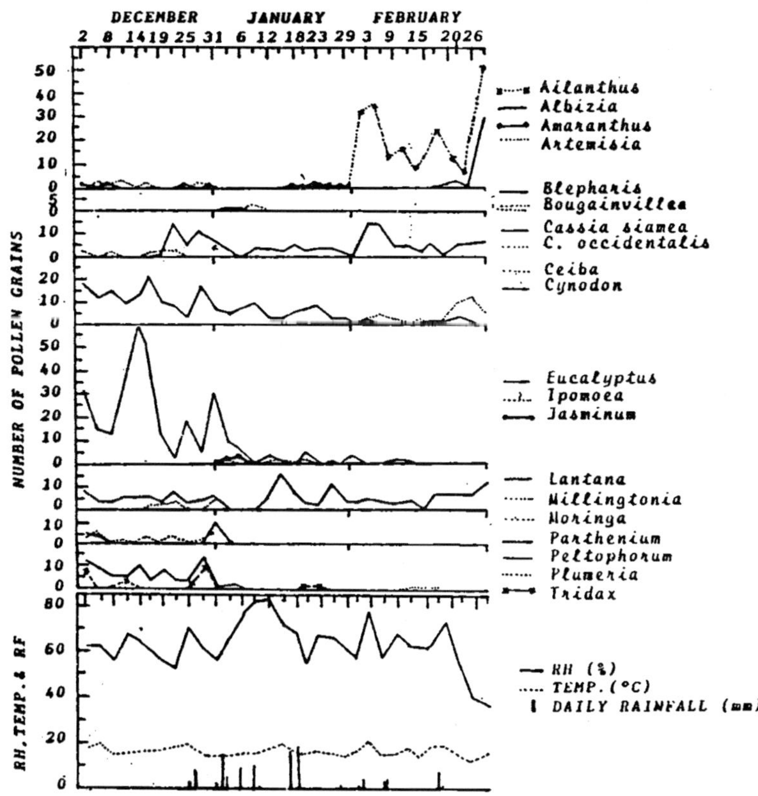

Fig. 16.1 : Pollen calendar and ombro-hygrothermic diagram of the winter : 1983-84

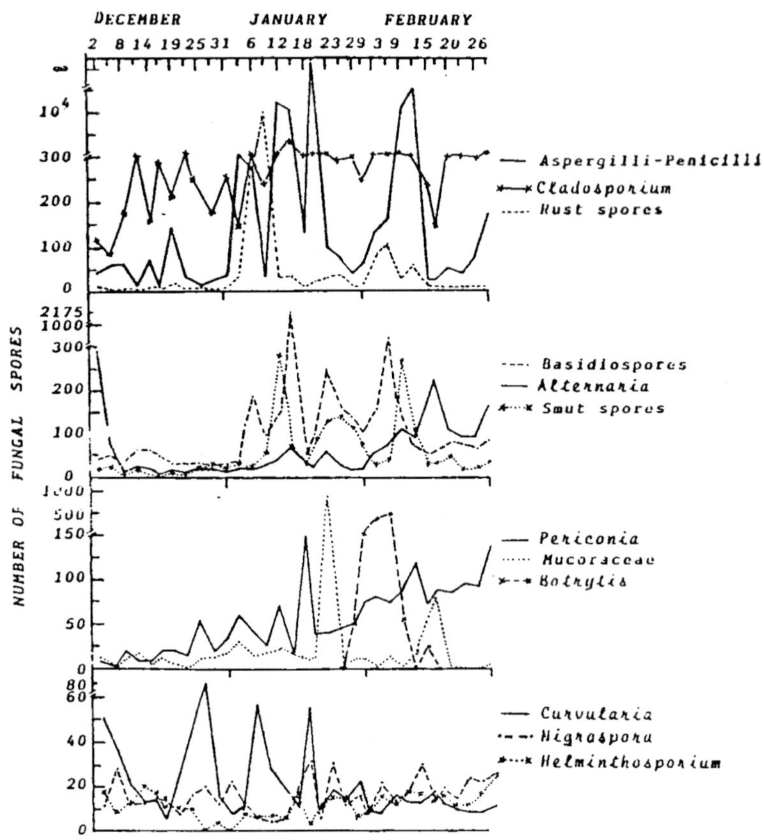

Fig. 16.2 : Biweekly fluctuation in the occurrence of certain dominant spore types in the winter : 1983-84.

high relative humidity was responsible for decrease in the atmospheric pollen load by accelerating the process of deposition.

On contrary to the pollen grains, the fungal spores showed maximum frequency during January followed by February and December. Thus, a significant positive correlations between atmospheric relative humidity and fungal flora is evident (Fig. 16.2). Similar relationship has also been established by several workers (Ramalingam, 1971; Bartzokas, 1975; Kumar, 1984). The Aspergilli group which may behave as allergens as well as pathogens, occurred in highest percentage (63.08%). The occasional trapping of entire conidiophore was responsible for huge quantity of these spores. *Cladosporium*, the universal dominant ranked second.

ACKNOWLEDGEMENT

I am greatly indebted to Dr. G.P. Agarwal, Emeritus Professor and to Professor M. Oommachan, Department of Biological Science, Rani Durgavati University, Jabalpur, for guidance and inspiration.

REFERENCES

1. Agarwal, G.P., Oommachan, M and Mishra, R.P. 1983. Aspects of environmental biopollution with special reference to pollen grains and fungal spores, Abst no. 2 *Nat. Conf, Environ.* Strategy, University of Kashmir and Srinagar, Srinagar (J&K).

2. Bartzokas, C.A. 1975. Relationship between the meteorolgical conditions and the airborne fungal flora of the Athens metropolitan area, *Mycopathologia 57 (1)* : 35-38.

3. Khare, K. 1984. *Air spora at Rani Durgavati Vishwavidyalaya campus with special reference to fungal spores*, M. Phil Dissertation, R.D. University Jabalpur (M.P.).

4. Kumar, Rajiv 1984. Studies on the aeromycospora of Dehra Dun city J. *Indian bot, Soc. 63* : 277-291.

5. Mishra, R.P. 1985. *Aerobiological studies with special reference to pollen grains and fungal spores of Jabalpur (M.P.)* Ph.D. Thesis, R.D. University, Jabalpur (M.P.).

6. Mishra, R.P. 1987. Studies on seasonal and diurnal variations in the occurrence of airborne spores of Basidiomycetes. In : *Perspectives in Mycological Research,* Today and Tomorrow's Printers and Publishers, New Delhi, Volume *1*: 243-252.

7. Ramalingam, A. 1971. Air spora of Mysore. *Proc, Indian Acad. Sci. 74* : 227-240.

8. Verma, K.S. and Khare, K. 1987a Study of air spora around Jabalpur University campus, *J. Econ. Tex Bot 1* : 87-90.

9. Verma, K.S. and Khare, K. 1987b. Fungal agents which damage paper material in library, in : *Atmospheric Biopollution* (ed. Chandra, N.) Environmental Publication KARAD, India pp. 193–197.

10. Verma, K.S. and Pandey, M. 1992. Airborne fungal spores in a hospital ward at Jabalpur, *Proc. Nat. Acad. Sci. India 62* (3) : 425-427.

11. Verma, K.S., Bhandari, S. and Chile, S. 1992. Volumetric analysis of air mycoflora of broiler and layer sheds, *Indian Journal of Poultry Science 27* (4) : 243-246.

INCIDENCE OF VIABLE AND CULTURABLE AIRMYCOFLORA IN JABALPUR

K.S. Verma and Karuna Shrivastava

ABSTRACT

The exposed plate method was used to trap airborne fungus spores in the three open locations of Jabalpur, India, for a period of one year. A total of 55 genera and 2499 fungal colonies were recorded and a distinct seasonal variation was observed in the more common ones.

On the exposed plates, spores of *Penicillium* and *Aspergillus* were more frequently trapped than any other genus. *Penicillium* (27.1%) was the main genus with *Aspergillus* (24.7%), *Fusarium* (8.9%), *Curvularia* (8.6%), *Trichoderma* (3.5%), *Rhizopus* (2.9%), *Alternaria* (2.8%), *Cladosporium* (2.2%), *Helminthosporium* (1.7%), *Scopulariopsis* (1.2%), *Periconia* (1.1%) and *Myrothecium* (1.1%) were also abundant. Other genera contributed 14.2% to the total airspora. Locality, environmental factors and vegetation appears to have influenced the spore type caught. Under favourable conditions some spore type attain high concentrations and on this account require consideration as possible allergens and plant pathogens.

INTRODUCTION

The atmosphere plays an important part in the transfer of propagules of many plant pathogenic fungi as well as those of saprophytic types. These airborne spores may serve as allergens when inhaled by susceptible persons. Thus, the nature of the airspora is of interest to allergists as well as plant pathologists.

In 1966, systematic studies were commenced to survey the live airborne spores of saprophytic microfungi, particularly those potentially able to colonize materials. Bruce (1963) incriminated the major fungal components of the airspora, including basidiospores.

The plant pathologists and allergists are primarily concerned with the concentration and composition of the airspora in relation to locality, season and weather elements. Many surveys have been carried out in which exposed gravity slide or nutrient plate techniques have been used and have furnished considerable data (Summarized in part by Feinberg, 1944, Maunsell, 1954). The routine sampling was done at the dense area and the market area to find variation in spore deposition in these different environments throughout the one year period by exposing nutrient petri-plates.

MATERIAL AND METHODS

For air sampling, exposed petriplate method was used. The method of trapping spores by petriplates exposure involves the principle of sedimentation under the force of gravity. Three sites were selected in the city at Jabalpur.

Site I (City area) - This is a residential colony have separate bungalows and apartments also surrounded by schools, colleges and play grounds.

Site II (Market area) - This is the city market have variety of shops and vegetable and fruit market.

Site III (Bhartipur area) - This is also a congested residential area of Jabalpur situated nearby market area.

Three petri-plates (9 cm. diameter) of freshly poured potato dextrose agar medium. (Agarwal and Hasija, 1986) were exposed at each site for five minutes at 16.30 hours twice in a month. Plates were exposed at one meter height from ground level at site II and III, while at 15 meter at site I. After exposure, the plates were incubated at 28 + 1°C and observed gradually for developing colonies, which were counted, examined and identified to generic and species level (Barnett and Hunter, 1972, Ellis 1971, 1976 and Subramanian 1971).

OBSERVATIONS

Table 17.1 gives the monthly occurrence of fungi at three sites of this sampling, whereas, Table 17.2 indicates occurrence of different fungi at three sites and their percentage contribution to the total colonies for the whole period.

Table 17.1 : Monthly concentration of fungal colonies during plate exposure

S. No.	Months	Site I	Site II	Site III	Total	Percentage
1.	January	26	51	31	108	4.32
2.	February	41	81	24	146	5.84
3.	March	89	113	96	298	11.92
4.	April	59	85	51	195	7.80
5.	May	83	69	62	214	8.56
6.	June	120	98	59	277	11.08
7.	July	30	54	37	121	4.84
8.	August	34	46	31	111	4.44
9.	September	69	73	36	178	7.12
10.	October	66	132	25	223	8.92
11.	November	172	107	85	364	14.56
12.	December	93	109	62	264	10.56
	Total	882	1018	599	2499	99.96

During the whole survey, 55 genera of fungi were recognized one yeast group and one indetermined sterile mycelium was also encountered. Some fungi which could not be identified, kept in a separate group 'indetermined fungi'.

A total of 2499 colonies were trapped one exposed petriplates of which 2407 were identified into 55 species of fungi, Ninety two colonies could not be identified thus kept under 'indetermined fungi'. Maximum number of colonies were recorded in the month of November followed by March, June, December and October. The maximum colonies were counted from the exposed petriplates of site II followed by site I and site III. Lesser number of colonies were encountered at the height of 15 meters than one meter. The two data of one meter height above ground level is significantly differ.

DISCUSSION

During the course of one year study as many as 35 genera were identified. The genus *Penicillium* was the predominant fungal genus in our studies accounting 27.12% of the total colonies. Two species of *Penicillium* was identified with a major season from July, August and October to December. The occurrence of this genus as a Winter type reported by Mishra (1972). Chakraborty (1976) reported it to be common from April to October and Sinha et al. (1982) reported its peak in May. *Aspergillus* formed the second major type (24.76%) with five species drawing maximum occurrence during April, May and November-December. *Aspergillus flavipes* (10.04%) was the commonest followed by *A. terreus* (6.60%) and *A. niger* (6.20%). *Aspergillus* was reported to occupy third place by Singh (1979), but Rati and Ramalingam (1976) reported it as most dominant type next to *Cladosporium* while Rajan et al. (1952) observed it as the most dominant one. *Fusarium* contributed (8.96%). Its presence in the airspora was reported by Sandhu et al. (1964), Agrawal et al. (1969), Rees (1964) and Calvo et al. (1980a).

Curvularia (8.60%) was also found in high frequency on exposed plates whereas *Trichoderma* (3.56%), *Rhizopus* (2.90%), *Alternaria* (2.84%) and *Cladosporium* (2.20%) was obtained in lower frequency, Sandhu et al. (1964) found *Alternaria* to be the most dominant (30.06%). The predominant genus *Cladosporium* was reported in comparatively less concentration (2.20%) on culture plates in this survey while Agrawal et al. (1969) found 28.4%, Rati and Ramalingam (1976) 70.25%, Sinha et al. (1982) 72.4%, Al-Doory et al. (1980) 20.4% and Çalvo et al. (1980a) 34.6%. *Cladosporium* has been shown to be the predominant type in most areas (Gregory and Hirst, 1957, Lecay, 1962) as well as other genera often encountered in exposed plate surveys include *Alternaria, Penicillium, Pullularia, Botrytis, Epicoccum* and *Aspergillus*, their relative importance varying with locality and climate.

The maximum colonies were counted from the exposed petriplates of site II (market area) at the height of 1 mt. above ground level followed by site I (city area) at 15 mt. above and site III (Bhartipur area) at 1 mt. above from ground level. Sampling was done at 16.30 hours by exposing the petriplates for five to ten minutes. In the market area, along with many open ditches, rotting edibles, other discarded materials, such as gunny bags, paper bags, packing materials, straw leaves and stems were the main herbingers of fungi. Due to the many ditches in this area, the market become very dirty.

Table 17.2 : Fungal spores trapped through culture plates exposure during 1985

S.No.	Name of Fungi	Site I No. of Colonies	%	Site II No. of Colonies	%	Site III No. of Colonies	%	Total No. of Colonies	Percentage of individual Colony
1	2	3	4	5	6	7	8	9	10
1.	*Acremonium strictum*	-	-	2	0.19	-	-	2	0.08
2.	*Acremonium* sp.	6	0.68	1	0.09	5	0.83	12	0.48
3.	*Alternaria alternata*	14	1.58	22	2.16	22	3.67	58	2.32
4.	*Alternaria* sp.	4	0.45	2	0.19	7	1.16	13	0.52
5.	*Aspergillus flavipes*	125	14.17	118	11.59	8	1.33	251	10.04
6.	*Aspergillus niger*	60	6.80	38	3.73	57	9.51	155	6.20
7.	*Aspergillus terreus*	28	3.17	109	10.70	28	4.67	165	6.60
8.	*Aspergillus sydowii*	-	-	-	-	2	0.03	2	0.08
9.	*Aspergillus ustus*	8	0.90	3	0.29	29	4.84	40	1.60
10.	*Aspergillus versicolor*	-	-	4	0.39	2	0.33	6	0.24
11.	*Beauvaria bassiana*	-	-	-	-	11	1.83	11	0.44
12.	*Bipolaris spicifera*	11	1.24	7	0.68	4	0.66	22	0.88
13.	*Botry displodia theobromae*	3	0.34	2	0.19	1	0.16	6	0.24
14.	*Chaetomium globosum*	6	0.68	8	0.78	4	0.66	18	0.70

1	*2*	*3*	*4*	*5*	*6*	*7*	*8*	*9*	*10*
15.	*Chaetomium murarum*	-	-	4	0.39	-	-	4	0.16
16.	*Chaetomium* sp.	-	-	1	0.09	-	-	1	0.04
17.	*Chrysosporium tropicum*	7	0.79	8	0.78	10	1.66	25	1.00
18.	*Chrysosporium* sp.	-	-	-	-	1	0.16	1	0.04
19.	*Cladosporium oxysporum*	23	2.60	12	1.17	13	2.17	48	1.92
20.	*Cladosporium* sp.	-	-	-	-	7	1.16	7	0.28
21.	*Cunnighamella echinulate*	3	0.34	-	-	1	0.16	4	0.16
22.	*Curvularia clavata*	31	3.51	65	6.38	46	7.67	142	5.68
23.	*Curvularia geniculata*	-	-	3	0.29	2	0.33	5	0.20
24.	*Curvularia lunata*	43	4.87	11	1.08	11	1.83	65	2.60
25.	*Curvularia prasadii*	-	-	3	0.29	-	-	3	0.12
26.	*Dictyoarthrinium* sp.	1	0.11	3	0.29	-	-	4	0.16
27.	*Drechslera specifera*	18	2.04	1	0.09	4	0.66	23	0.92
28.	*Emericella nidulans*	5	0.56	16	1.57	3	0.50	24	0.96
29.	*Eupenicillium shearii*	-	-	2	0.19	-	-	2	0.08
30.	*Fusarium equiseti*	6	0.68	9	0.88	8	1.33	23	0.92
31.	*Fusarium moniliforme*	4	0.45	22	2.16	22	3.67	42	1.92

1	2	*3*	*4*	*5*	*6*	*7*	*8*	*9*	*10*
32.	*Fusarium oxysporum*	131	14.85	11	1.08	10	1.66	152	6.08
33.	*Fusarium* sp.	-	-	-	-	1	0.16	1	0.04
34.	*Haplosporella* sp.	-	-	-	-	2	0.33	2	0.08
35.	*Helminthosporium anomalus*	6	0.68	6	0.58	6	1.00	18	0.72
36.	*Helminthosporium sativum*	18	2.04	7	0.68	1	0.16	26	1.04
37.	*Humicola* sp.	12	1.36	12	1.17	3	0.50	27	1.08
38.	*Keratinophyton terreum*	1	0.11	-	-	-	-	1	0.04
39.	*Mucor* sp.	1	0.11	-	-	-	-	1	0.04
40.	*Myrothecium verrucaria*	15	1.70	9	0.88	5	0.83	29	1.16
41.	*Nigrospora sphaerica*	4	0.45	11	1.08	9	1.50	24	0.96
42.	*Paecilomyces fumorescus*	-	-	2	0.19	4	0.66	6	0.24
43.	*Paecilomyces variotii*	-	-	4	0.39	-	-	4	0.16
44.	*Penicillium stipitatum*	141	15.98	277	27.21	88	14.69	506	20.24
45.	*Penicillium citrinum*	64	7.25	53	5.20	55	9.18	172	6.88
46.	*Periconia* sp.	4	0.45	14	1.37	11	1.83	29	1.16
47.	*Phoma* sp.	1	0.1	-	-	-	-	1	0.04
48.	*Rhizopus* sp.	29	3.28	27	2.65	17	2.83	73	2.92

1	2	3	4	5	6	7	8	9	10
49.	*Schizophyllum commune*	-	-	3	0.29	-	-	3	0.12
50.	*Scopulariopsis brevicaulis*	2	0.22	14	1.37	14	2.33	10	1.20
51.	*Sporotrichum pruinosum*	-	-	2	0.19	3	0.50	5	0.20
52.	*Talaromyces trachyspermus*	-	-	5	0.49	1	0.16	6	0.24
53.	*Thielavia terricola*	1	0.11	1	0.09	1	0.16	3	0.12
54.	*Trichoderma viride*	23	2.60	42	4.12	24	4.00	89	3.56
55.	*Trichothecium roseum*	3	0.34	4	0.39	-	-	7	0.28
56.	Indetermined sterile mycelium	-	-	-	-	1	0.16	1	0.04
57.	Yeast group	-	-	-	-	1	0.16	1	0.04
58.	Indetermined fungi	20	2.26	38	3.73	34	5.67	92	3.68

Yeast colony and sterile mycelium also reported on culture plates. Rees (1964) recorded high incidence of yeast (12.2%) and nonsporulating fungi (7.4%). Chatterjee (1987) also recorded yeast and mycelia sterila on culture plate studies at Bangalore.

One of the major advantage of culture plate studies is several important types such as *Aspergillus, Penicillium, Fusarium*, etc., which constitutes major percentage of total colonies formed a lesser percentage of the catches when trapped by suction sampler or often went unrecognized by virtue of their similar morphological characters. The other types which were obtained on culture plates were *Mucor, Cunninghamella, Beauvaria, Bipolaris, Botryodiplodia, Chaetomium, Chrysosporium, Dictyoarthrinium, Drechslera, Haplosporella, Helminthosporium, Myrothecium, Nigrospora, Paecilomyces, Periconia, Phoma, Schizophylium, Scopulariopsis, Sporotrichum, Thielavia*, and *Trichothecium*.

Some fungi were also trapped in their perfect stage such as *Emericella, Eupenicillium, Keratinophyton*, and *Talaromyces*.

Spore viability is an important factor for colony formation Pathak and Pady (1965) found that an average of 42% of *Cladosporium* spores were germinable while 62% was reported by Pady and Gregory (1963). Harvey (1967) stated that viability is also diurnally variable and dependent on the clumping degree of the spores. In general, maximum number of colonies were encountered during November followed by March, December and June which were the dry and humid months with moderate temperature which favoured the growth of fungi and their release in the atmosphere.

The ultimate aim of this work is to have a permanent systematic record of spore types occurrence in the atmosphere during different varying conditions. This record further helps to clinicians in the treatment of fungal allergy and also in taking precautions in plant disease forecasting to agriculturists.

ACKNOWLEDGEMENTS

The authors are grateful to Dr. S.K. Hasija, Professor and Head, Department of Biological Sciences, R.D.V.V., Jabalpur and Dr. G.P. Agrawal, Emeritus Scientist for their encouragement and valuable suggestions.

REFERENCES

1. Agarwal, G.P. and Hasija, S.K. 1986. Microorganisms in the laboratory. Print House (India) Lucknow pp. 155.

2. Agarwal, M.K., Shivpuri, D.N., and Mukherjee, K.G., 1969. Studies on the allergenic fungal spores of the Delhi India Metropolitan area J. Allergy 44: (4) 193-203.

3. Al-Doory, Y.J., Domson, F., Howard, W.H. and Sly, R.M. 1980. Airborne fungi and pollen of the Washington D C Metropolitan area Ann. Allergy 45 : 360-367.

4. Barnett, H.L. and Hunter, Barry. B. 1972. Illustrated genera of imperfect fungi Burgess Publishing Company Minneapolis Minnesota.

5. Bruce, R.A., 1963. Bronchial and skin sensitivity in Asthma Int. Arch. Allergy Basel 22 : 294-305.

6. Calvo, M.A., Guarro, J. Suarez, G and Ramirez, C. 1980a. Airborne fungi in Barcelona city (Spain) I A two year study 1976-1978 Mycopath. 71 : 89-93.

7. Chakraborty, R. 1976. Aeromycology of the Indian Botanic garden Calcutta Asp. Allergy and Appl. Immunol 9 : 151-157.

8. Chatterjee, M. 1987. Aerobiological studies of Banglore north with particular reference to fungal aeroallergens Ph.D. Thesis Banglore University Jhana Bharathi Banglore.

9. Ellis, M.B. 1971. Dematiaceous Hyphomycetes Common Wealth Mycological Institute Kew Surrey England.

10. Ellis, M.B. 1976. More dematiaceous Hyphomycetes Common Wealth Mycological Institute Kew Surrey England.

11. Feinberg, S.M. 1944. Allergy in Practice year book Publishers Chicago.

12. Gregory, P.H. and Hirst, J.M. 1957. The summer air Spore at Rothamsted in 1952 J. Gen. Microbiol 17 : 135-152.

13. Harvey, R. 1967. Airspora studies at Cardiff I *Cladosporium* Trans. Br. Mycol. Soc. 50 : 479-495.

14. Lacey, M.E. 1962. Summer airspora of two contrasting adjacent rural sites J. Gen. Microbiol. 29 : 485-501.

15. Mishra, R.R. 1972. Aeromycology of Gorakhpur IV Periodical fluctuations of airspora Mycopath. Mycol. Appl. 48 : (2 & 3) 213-222.

16. Mounsell, K. 1954. Respiratory allergy to fungus Spores Progs. Allergy 4: 457-520.

17. Pady, S.M. and Gregory P.H. 1963. Numbers and Viability of airborne hyphal fragments Trans. Br. Mycol. Soc. 46 : (4) 609-613.

18. Pathak, V.K. and Pady, S.M. 1965. Numbers and Viability of certain airborne fungus Spores Mycologia 57 : 301-310.

19. Rajan, B.S.V., Nigam, S.S. and Shukla, R.K. 1952. A study of the atmospheric fungal flora at Kanpur Proc. Ind. Aced. Sci. 35 : (B) 33-37.

20. Rati, E. and Ramalingam, A. 1976. Airborne Aspergilli at Mysore Asp. Allergy and Appl. Immunol. 9 : 139-149.

21. Rees, R.S. 1964. The airspora of Brisbane Austr. J. Bot. 12 : 185-204.

22. Sandhu, D.K. Shivpuri, D.N. and Sandhu, R.K. 1964. Studies on the airborne fungal spores in Delhi - their role in respiratory allergy Ann. Allergy 22 : 374-384.

23. Singh, N.I. 1979. Seasonal germination Potential of airborne Aspergillus - A common allergen of shillong and its suburbs J. Palynol. 15 : (2) 85-90.

24. Sinha, S. Chakraborty, R. and Mishra, A.K. 1982. Seasonal changes in 5 common constituents of the airspora of Calcutta Trans. Bose. Res. Inst. 45: (3 & 4) 109-114.

25. Subramanian, C.V. 1971. Hyphomycetes (New Delhi) ICMR, India.

PALYNOFOSSILS OF MIDDLE AND UPPER GONDWANAS IN SATPURA BASIN–PALAEOENVIRONMENTAL SIGNIFICANCE

Parmod Kumar

INTRODUCTION

The sedimentary sequences of the Satpura Basin largely deposited under fresh water continental fluviatile of lacustrine environments, were designated as Gondwana and were divided into Lower, Middle and Upper Gondwanas under the tripartite classification, as proposed by *Feistmantel* (1876a,b) followed by Vredenburg (1910) and Lele (1962, 1964), which were equivalents to the Permian, Triassic and Jurassic systems of Europe. They are characterised by the *Glossopteris*, *Dicroidium* and *Ptilophyllum* floras, respectively (Krishnan, 1982). The term Middle Gondwana was also supported by the intervention of an arid or semiarid continental deposits containing Triassic reptiles and amphibian faunas, between the normal conditions of Lower and Upper units. The Middle Gondwana in Satpura basin had a thick deposit of the Pachmarhi ferruginous sandstones (about 750 m), Denwa red clays (about 350 m) and Bagra conglomerates (about 250 m) of the Mahadeva Formation. The Upper Gondwana was represented by the Jabalpur Formation (about 100-150 m) which mainly comprised ferruginous sandstones, white or coloured clays, carbonaceous shales and thin coaly layers (Crookshank 1936; Sastry *et al.*, 1977) (Map 18.1). These lithological, megafloral and faunal variations might have been the result of the environmental variations. Environmental is a very complex phenomenon. In addition to its living components it has a number of climatic variables, viz., temperature, pressure, humidity, evapotranspiration, altitudes, latitudes, land-sea distribution, ocean and wind currents, etc. The role of these parameters are very difficult to determine independently because their combined effect plays role in shaping the climate and there by the flora and fauna geological past.

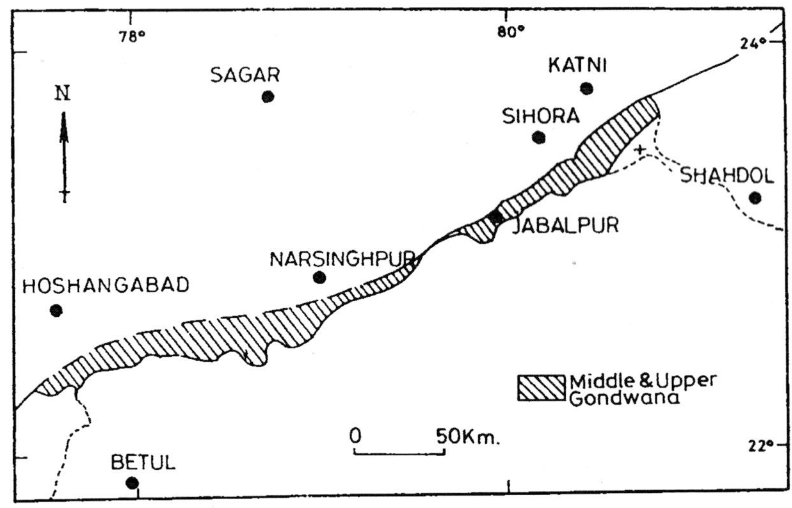

Map 18.1 : Showing distribution of Middle and Upper Gondwanas in Satpura basin, M.P.

A seminar was organised on climatic vicissitudes during Gondwana of India, for considering various evidences from palaeogeography (Bharadwaj, 1976); sedimentological variations (Laksar and Mitra, 1976), (Dutta, 1976); Mineralogy (Singh, I.B., 1976); fauna (Shah, 1976); flora (Vakhrameev, 1965; Gupta, 1966; Lele, 1962, 1964, 1976) and palynoflora (Kar, 1976). Srivastava (1978) suggested palaeoclimate on the basis of plants and palynospores relationships during Cretaceous Period. Recently, Ramanujam (1983, 1993), Kumar (1992, 1993 in Press b) have dealt the palaeoclimate on the same pattern. These efforts although have direct bearing on palaeoecology, are indirectly supporting the climatic conditions.

In the present communication the author intends to decipher the impacts of climate on the spore pollen characters during the Middle an Upper Gondwana times.

For this one should know the correct morphographical features of the pollen-spores occurring in different palynoassemblages which are recorded from time to time from different outcrop sections or subsurfaces.

(i) ***Palynological assemblage from Middle Gondwana***

(i) (a) *Falcisporites-Alisporites-Satsangisaccites Assemblage zone II at 200 m depth*

(b) *Staurosaccites-Lunatisporites-Lueckisporites Assemblage zone I at 100.5 m depth*

Locality : Bore-hole ANH-1 at Anhoni 22°38 : 78°21')near Jhirpa, Pachmarhi area, M.P.

Formation : Denwa Formation (inclu. red clays, buff and carbonaceous shales). Middle Gondwana.

Age : Lower-Middle Triassic age.

Nandi (1991) reported dominance of gymnospermic nonstriated disaccate pollen e.g. *Falcisporites, Alisporites, Klausipollenites, Satsangisaccites, Nidipollenites*, etc. Vijaya (pers. communication) apart from these forms some more taeneate forms *Lunatisporites, Lueckisporites, Infernopollenites*, etc. were also present in Nandi's spore-pollen slides. Some pteridophytic trilete cavate/zonate spores were identified, i.e. *Playfordiaspora, Densoisporites, Lundbladispora*, etc. and other sculptured or structured trilete spores as *Callumispora, Leschikisporis, Converrucosisporites, Osmundacidites*, etc. Bryophytic spores were not reported so far, (Table 18.1).

Ephedralean elator bearing pollen *Equisetosporites* and trilete structured spore *Punctatisporites, Retusotriletes* (Figs. 4,5,6, Nandi, 1991) need a thorough reidentification.

(c) *Falcisporites-Satsangisaccites-Alisporites Palynoassemblage-*

Locality : Chota Mahadeva scrap

Tamia cliff (22° 21' : 78° 40')

Tamia, Chhindwara district, M.P.

Formation : Pachmarhi Formation (basal buff coloured sandy shales), Mahadeva Group, Middle Gondwana.

Age : Lower Triassic.

Kumar (C., in Press) recovered palynofossils which are as stated below : Dominant forms (20-30% or above) are : *Falcisporites, Satsangisaccites, Alisporites.* Subdominant (10-20%) components of the assemblage are *Striatopodocarpites, Scheuringipollenites, Densipollenites.* The common (5-10%) forms indicated by *striatites, Lunatisporites, Corisaccites, Klausipollenites, Choradasporites.* The components represented as Fair (2-5%) are * *Goubinispora, Distriatites,* * *Nidipollenites.* The poor/rare (0.5-2%) forms are *Weylandites,* * *Playfordiaspora, Callumispora,* * *Osmundacidites,* etc.

The assemblage is mostly characterized by the newly incoming elements as dominance of nonstriated disaccates with representation of Fair/Rare forms (*) indicate a change in the climatic pattern. Pteridospermic and gymnospermic elements are forming the major bulk of the assemblage whereas cavate/cingulate, plicates and trilete are very scanty.

(ii) ***Callialasporites-Araucariacites-Podocarpidites palynoassemblage***

Localities - Kotri (22° 41' : 78° 50'), Hathidoba (22° 43' : 78° 54'), Napupura (22° 20' : 77° 36'), Khardi nadi (22° 24' : 77° 36'), in Jatamao region; Khatama caves (22° 30' : 77° 45'), Morghat (22° 28' : 77° 42'). All are outcrop sections.

Formation - Jabalpur Formation (Chaugaon Bed-Carbonaceous shales), Satpura basin, M.P.

Age - Upper Jurassic, Upper Gondwana.

Table 18.1 : Palynomorphs with its climate during Middle and upper gondwana in satpura basin

Geological Age	Triassic	Lower-Mid DLE Jurassic	Upper Jurassic	Early Cretaceous				
Flora	Dicroidium	Ptilophyllum		Flora				
Palynozone	Falcisporites	Podocarpidites	Calliala sporites	Arauca riacites	Calliala sporites +T. Reticulatus	Podosporites	Chief characters	Habitats
Sections/Areas	Anhoni (22° 38' : 78° 21') & Tamia cliff (22° 21' : 78° 40')	Parsapani (78° 2' : 22° 36')	Morghat (77° 26' : 22° 22' Kotri(78° 54' -58° : 22° 39' -43')	Schora (79° 20' : 22° 52') Hathnapur (79° 3' : 22° 45')	Ellichpur (79° 20' : 21° 22')	Pat-Baba Ridge (79° 60' : 23° 12')		
Climate	Warm Temperate, dry with restricted rain	warm dry with less humidity	moderate climate with better humidity	Tropical to Subtropical good rains with swampy water logged conditions with seasonal variations				
Palynoforms	%	%	%	%	%	%		
Class-Gymnospermae								
Order-Cycadeles								
Genera :								
Cycadopites	-	3.5	2.0-3.0	16.0-18.0 (S.D.)	1.0	1.0	Spindle shaped monosulcate, smooth	Tropical/subtropical region xerophytic + need water for completing its lifecycle (Moist soil)

1	2	3	4	5	6	7	8	9
Monosulcites	-	-	-	+	-	-	Elongated rounded, monosulcate, smooth	
Order-Coniferales								
Genera :								
Alisporites	10-20.0	9.0	+	+	-	-	Nonstriated-Disaccates, Haploxylonoid	Cool, hilly places of upland areas
Klausipollenites	5-10.0	-	-	-	-	-	" "	Upland areas
Falcisporites	20-30.0						" "	" "
Family-Podocarpaceae								
Genera :								
Podocarpidites	+	36.0 (D)	4.0-5.0	6.0-8.0	5.0-7.0	3.5	Disaccate, diploxylonoid	Southern hemisphere in warm hilly upland places.
Callialasporites	-	17.0 (S.D.)	35.0-40.0 (D)	10.0-12.0	50.0(D)	1.0	Subsaccate smooth/ornamented pollen	" "
Podosporites	-	12.0	+	1.0-	1.0	46.0		" "
Family-Araucariaceae								
Genera :								

1	2	3	4	5	6	7	8	9
Araucariacites	-	16.0 (S.D.)	22.0-25.0 (S.D.)	30.0-32.0 (D)	5.0	3.0	Nonsaccate, inaperturate or ulcate, ornamented (granulose)	Southern hemisphere, cool, hilly upland areas.
Family-Cheirolepidaceae								
Genera :								
Classopollis	-	1.0	1.0	2.0-4.0	-	-	Operculate, intrabaculate	Near the margin of basin
Gliscopollis	-	-	-	+	-	-	Operculate	" "
Class-Bryophyta								
Genera :								
Aequitriradites	-	-	-	+	-	-	Zonate, hilate	Tropical, Moist places
Coptospora		-	-	+	-	-	nonzonate, hilate	woodlands or near the banks of stream or shaded cliffs or swampy meadows
Cooksonites		-	-	+	+	-	Cingulate, hilate	wet open.
Triporoletes (Rouseisporites)	-	-	-	+	2.0	-	Zonate distally muroid perinate	Wet places
Crybelosporites		-	-	+	-	-	Trilete, perinate, cavate	Warm/dry

1	2	3	4	5	6	7	8	9
Stereisporites	-	-	-	+	-	-	Azonate, trilete distally polar region thicker	Wet places
Class-Pterido-phyta								
Family-Lycopo-diaceae								
Genera :								
Lycopodiums-porites	-	-	+	+	-	-	Trilete, Reticulate	Tropical/subtropical forests heaths, humus soils in moist shady places
Sestrosporites	-	-	-	+	-	-	Trilete, crassate	
Family- Selagin-ellaceae								
Genera :								
Neoraistrickia	-	-	-	+	-	-	Trilete, Apiculate	(xerophytic) grow on warm and dry soil
Densoisporites	+	+	+	+	-	-	Zonate, perinate Trilete, sculptine ornamented	continentality grow on warm, dry soil
Lundbladispora	+	-	-	-	-	-	" "	continentality, grow on warm, dry soil

1	2	*3*	*4*	*5*	*6*	*7*	*8*	*9*
Playfordispora	+	-	-	-	-	-	Zonate, sculptine OL Patterned	" "
Order-Filicales								
Family-Gleicheni-aceae								
Genera:								
Gleicheniidites	-	-	-	+	-	-	Trilete, Crassi-tudes, smooth	Tropical subtropical shady and moist places both in hills and in plains of thick forests.
Peregrinisporites	-	-	-	+	-	-	Trilete, Crassi-tudes, Apiculate	
Concavisporites	-	-	+	+	-	-	Trilete, Kyrtome, smooth	" "
Family-Osmund-aceae								
Genera								
Osmundacidites	+	+	+	+	-	-	Trilete, Apicu-late	Tropical and temp-erate shaded and cooly regions
Baculatisporites	-	-	+	+	-	-	Trilete, Apicu-late	" "
Biformaesporites	-	-	+	+	-	-	"	" "

1	2	3	4	5	6	7	8	9
Todisporites	-	-	+	1.0	-	1.0	Trilete, smooth	" "
Genera:								
Matonisporites	-	-	+	+	+	-	Trilete, Auricu-late	
Dictyophllidites	-	+	+	+	+	-	Trilete, crassate	
Boseisporites	-	+	+	+	+		Trilete, Auricu-late punctate.	+ Tropical, less rains with xeric conditions
Lametatriletes	-	+	+	+	+	-	Trilete, Auricu-late Ornamented (punctate)	
Callispora	-	1.0	+	+	+	10.0	" "	
Family-Hymeno-phllaceae								
Genus:								
Biretisporites	-	-		+	-		Trilete, smooth	Tropical, rain forest
Family-Cyatheaceae/ Dicksniaceae								
Genera:								
Cyathidities	-	1.0	1.0	2.0-4.0	1.0	22.0 (S.D.)	Trilete, smooth	" "
Haradisporites	-	1.0	1.0	1.0-2.0	-	-	Trilete, smooth/ ornamented	" "

1	*2*	*3*	*4*	*5*	*6*	*7*	*8*	*9*
Family-Schizaeaceae								
Genera:								
Ischyosporites	-	-	-	+	1.0	-	Trilete, Fovereticulate, Auriculate	Tropical moist and shaded places
Klukisporites	-	+	+	1.0	-	-	Trilete, Foveoreticulate	" "
Trilites	-	-	+	+	-	-	Trilete, Foveoreticulate/ apiculate, Auriculate	Warm, dry less rains
Cicatricosisporites	-	-	+	+	-	-	Trilete, striations/Ribbed (costate)	Tropical warm, wet land
Family-Polypodiaceae								
Genera:								
Monolites	-	-	+	+	1.0	-	Monolete, smooth, wall thick.	Tropical/subtropical hill tops (1500 to 2700 m height
Metamonoletes	-	-	-	-	-		Monolete, Apiculate	
Dettmannites	-	-	+	+	-	-	Monolete, Apiculate	Warm conditions

1	2	3	4	5	6	7	8	9
Laevigatosporites	-	-	-	+	-	-	Monolete, smooth	
Incertae sedis:								
Retusotriletes	+	-	-	-	-	-	Trilete, smooth	Tropical to sub-tropical warm humid
Punctatisporites	+	-	-	-	-	-	Trilete, punctate	
Equisetosporites	+	-	-	-	-	-	Two tightly appressed elators with the body	

Note - D = dominance ; S.D. = subdominance ; (+) = present ; (-) = absent.

Kumar and Kulshreshtha (1979), Kumar (1992, 1993 in press-a), Maheshwari and Kumar (1979) reported dominance of gymnospermic *Callialasporites* followed by *Araucariacites* and fair amount of *Podocarpidites* some characteristic Lycopodiales and selaginellid spores, viz.,*Lycopodiumsporites* and*Densoisporites* are present in the assemblage. Twenty pteridophytic spores are present here which are given in Table 18.1. These spores are present as poor elements in this palynoassemblage as known from different outcrop sections.

(iii) *Araucariacites-Callialasporites-Podocarpidites Palynoassemblage*

Localities : Sehora (22° 52' : 79° 20'), Hathnapur (22° 45' : 79° 3') and Lameta Ghat (23° 6' : 79° 50') outcrop sections;

Formation : Jabalpur Formation (Carbonaceous shales and thin coaly layers).

Age : Early Cretaceous (Neocomian), Upper Gondwana.

Singh (1966), Singh and Kumar (1966), Bharadwaj *et al. (1972)* (1972) and Kumar (1973) described gymnospermic pollen genera among which dominance of Araucariacites followed by *Callialasporites* had been reported. The Lycopodiales, Selaginallales are poor whereas 24 pteridophytic spore genera are well reported (Table 18.1). The assemblage contained six bryophytic spores, viz., *Aequitriradites, Cooksonites, Triporoletes (Rouseisporites)*, etc. also given in Table 18.1.

(iv) *Callialasporites-Triporoletes reticulatus palynoasseemblage*

Locality - Ellichpur (77° 32' : 21° 22') outcrcp section, Gawilgarh hills, Maharashtra;

Formation - Jabalpur Formation (Carbonaceous shales);

Age - Early Cretaceous (Neocomian-Aptian), Upper Gondwana.

Kumar(1992) recorded dominance of gymnospermic pollen *Callialasporites* followed by fair amount of *Araucariacites* and *Podocarpidites* (Table 18.1). Five gymnospermic pollen genera are recorded. Three bryophytic, and 9 pteridophytic spore genera were reported (Table 18.1). Bryophytic *Triporoletes reticulatus* is a unique occurrence in this palynoassemblage but poor in percentage frequency.

(v) *Podosporites-Cyathidites-Callispora palynoassemblage*

Locality - Patbaba Ridge (79° 50' : 23° 12') subsurface.

Formation - Jabalpur Formation (carbonaceous shales);

Age - Early Cretaceous (Neocomian) Upper Gondwana.

Kumar (1986) reported a palynoassemblage which and gymnospermic triwinged pollen *Podosporites* (equivalent to *Microcachryidites* Balme 1957 from IIb Microflora in Western Australia) as a dominant member. Five gymnospermic pollen were recorded. It is followed by only three pteridophytic spores, viz., *Cyathidites* (22.0%) and *Callispora* (10.0%). The details of other elements are also given in Table 18.1.

These pollen from the above palynoassemblages possess certain significant morphographical exinal features which are described in detail here as below.

(A) (i) *Nonstriated haploxylonoid disaccate bearing pollen*

The gymnosperm pollen, viz., *Falcisporites, Alisporites, Klausipollenites* (Fig.15), *Satsangisaccites*, etc. are found during Triassic period of Middle Gondwana as prominent members of the palynosassemblage. A few taeneate disaccates with haploxylonoid (sacci shroter or equal in vertical length in relation to central body) saccate pollen, viz., *Lunatisporites*, are also present in the assemblage.

(ii) *Nonstriated diploxylonoid disaccates bearing pollen*

The Lower-Middle Jurassic, and Lower Cretaceous had non-striated disaccates having diploxylonoid (sacci greater in vertical length in relation to central body) saccus, e.g. *Podocarpidites* (Fig. 14) dominant in former, whereas, polysaccates, *Podosporites* (Figs. 17, 18) (= *Microcachryidites*), dominants in the latter assemblage.

(iii) *Subsaccate or zonate/flange with distally ulcate pollen*

Subsaccate equatorially mono-trinotched, vesiculate/frilled (rimula) extension of exoexine behave as cavate air filled structures devices for floating in the air, as in *Callialasporites* (Figs. 19-21) dominated the Upper Jurassic and in some patches of the Lower Cretaceous palynoassemblages.

(iv) *Nonsaccate, Inaperturate with distally ulcate pollen*

Araucarian pollen *Araucariacites* (Fig. 16) is characterised by nonsaccate inaperturate with distal polar subcircular thinner exinal area (ulcus). It does not provide any modifications over the exine. It may be envisaged that cool, temperate, climate was existing around the upland areas during Lower Cretaceous in some of the patches.

In support of these suppositions Courtinat (1987) performed an experiment and observed that pollen having non saccus *Araucaria excelsa* R. Brown (similar to fossial pollen *Araucariacites* Cookson 1947) changed into saccus bearing pollen *Callialasporites monoalasporus* Dev. *C. segmentatus* (Balme) Dev and *C. trilobatus* (Balme) Dev. When treated under warmer conditions.

Thus the occurrence of disaccates, polysaccates, subsaccates (Podocarpaceous) and nonsaccate (Araucarian) pollen indicate that dry, warm/cool, temperate conditions with less rains and seasonal variations were prevailing at that upland high altitude regions where such plants were growing well (Table 18.2).

(B) The other characteristic morphographical features are developed in the bryophytic and pteridophytic spores. To continue and to complete their life cycles in a relatively harsh environment these spores modify their protective coverings i.e., exine. These exinal modifications help in protecting the living inner genetical components (Necleus, Cytaplasm, etc.) from the unfavourable climatic conditions by checking the evapotranspiration and save them from the microbial activities, etc. It also facilitate some more significant activities, e.g. Harmomegathy, imbibing of water from the atmosphere, and balancing the water composition of the living genetic components.

The Modifications are described here as under :

1. *Cingulate morphographic character of spores*

(i) *Cingulate spores*

The extra exinal layerings or thickenings (Cingulum) are developed at the equatorial regions of the spore body, viz., *Murospora*(Fig. 10) and *Cooksonite* (Table 18.2).

Table 18.2 : Palynomorphs with chief, characters and their probable affinities

Palynomorphs	Chief characters	Probable Affinities
Anteturma-*SPORITES* H. Potonie' 1893		
Infraturma-*Laevigati* (Bennie & Kidston) Pot. 1956		
Genera:		
Cyathidites Couper 1953	Trilete, Acavate, Azonate Triangular, Laevigate exine	Cyatheaceous/Dicksoniaceous ferns viz. *Coniopteris* spp.
Stereisporites Pflug 1953	Trilete, Acavate, Azonate, Triangular, Laevigate exine with thickenings in radial equatorial and distal polar regions.	Sphagnaceae: *(Sphagnum)*
Dictyophyllidites (Couper) Dettmann 1963	Trilete, Acavate, Azonate, Laevigate with thickenings about the Trilete margins	Matoniaceae : *Phlebopteris, Matonidium*
Infraturma-*Apiculati* (Bennie & Kidston) Pot. 1956		
Genera :		
Concavissimisporites Delc. & Sprum.) Dettm. 1963	Trilete, Acavate, Azonate, Triangular, Distally verrucate exine.	Cyatheaceae : *Cyathea, Dicksonia Lygodium*

Palynomorphs	Chief characters	Probable Affinities
Osmundacidites Couper 1953	Trilete, Acavate, Azonate, Spherical/ circular, exine Apiculate (granulate)	Osmundaceae : *Osmundopsis, Todites, Osmunda* (living form)
Baculatisporites Thomson & pflug 1953	Trilete, Acavate, Azonate, Circular, Apiculate (baculae, setulae/clavae) exine.	Osmundaceae : *Osmunda* spp.
Neoraistricki Potonie' em. Bharad. & Kumar 1972	Trilete, Acavate, Azonate, Roundly triangular, Apiculate (baculate at both faces) exine.	Selaginellaceae : *Selaginella.* spp.
Ceratosporites Cookson & Dettmann 1958	Trilete, Acavate, Subtriangular, Apiculate, (distallyclavate, setullate, spinulate) exine.	Selaginellaceae : *Selaginella*
Infraturma-*Murornati* Potonie's Kremp 1954		
Genera :		
Lycopodiumsporites Thiergart ex elcourt & Sprumont 1955	Trilete, Azonate, Acavate, distally and equatorially reticulate exine.	Lycopodiaceae : *Lycopodium* spp.
Klukisporites Couper 1958	Trilete, Azonate, Acavate, Tirangular, distally reticulate (foveoreticulate) exine	Schizaeaceae : *Klukia* spp. or *Stachypteris* spp.

Palynomorphs	Chief characters	Probable Affinities
Cicatricosisporites Potonie' & Gelletiche 1933	Trilete, Azonate, Acavate, Triangular exine with three series of muri at proximodistal and each equatorial region.	Schizaeaceae : *Anemia* spp.
Infraturma - *Auriculati* Schopf em. Dettm. 1963		
Genera :		
Matonisporites Couper em. Bharad. & Kumar 1972	Trilete, Acavate, Zonate, Auriculate, (differentially thickened exine at equatorial radial regions)	Matoniaceae : *Phlebopteris* spp. or *Dicksonia* spp.
Callispora Dev em. Bharad. & Kumar 1972	Trilete, Triangular, Acavate, Zonate, Auriculate (differentially inciplently thickened) punctate exine.	Matoniaceae
Boseisporites Dev em. Bharad. & Kumar 1972	Trilete, Triangular, Acavate, Zonate (Auriculate), punctate exine.	Matoniaceae
Lametatriletes Singh & Kumar 1973	Trilete, Triangular, Auriculate, punctate exine.	Matoniaceae
Trilites Erdtm. ex copuer em. Dettm. 1963	Trilete, Avacate, Zonate, Auriculate (differentially thickened in equatorial,	*Dicksonia* spp.

Palynomorphs	*Chief characters*	*Probable Affinities*
	radial regions) and Apiculate (Verruco-rugulate) exine.	
Ischyosporites Balme 1957	Trilete, Acavate, Zonate, Auriculate (distally and equatorially foveo-reticulate exine).	Schizaeaceae
Infraturma-*Tricrassati* Dettmann 1963		
Genera :		
Gleicheniidites (Ross ex Delc. & sprum) Dettm. 1963)	Trilete, Acavate, Zonate (equatorial; interradial exinal thickenings i.e. Crassitudes) exine.	Gleicheneaceae : *Gleichenia* spp.
Sestrosporites Dettmann 1963	Trilete, Acavate, Exine differentially thickened in equatorial, interradial regions (Crassitudes) Apiculate (foveo-reticulate).	Spores of *Lycopodium* spp.
Infraturma-*Cingulati* Potonie' & Klaus em. Dettm. 1963		
Genera :		

Palynomorphs	Chief characters	Probable Affinities
Murospora Somers 1952	Trilete, Acavate, zonate (cingulate) Triangular, differentially thickened, exine, thicker equatorially.	Dicksoniaceae : *Dicksonia* spp.
Contignisporites Dettmann 1963	Trilete, Acavate, zonates (cingulate) distally murornate (costate)	? *Anemia* spp. not definitely known
Suprasubturma-*Perinotriletes* Erdtm. em. Dettm. 1963		
Genera :		
Crybelosporites Dettmann 1963	Trilete, cavate, perinate, sculptine, reticulate, surface OL pattern (foveolate or Foveoreticulate) exine	Marsiliaceae : spores of *Pilularia* spp. *Marsilea* spp. *Reguellidium* spp.
Densoisporites (Weyl. & Krieg, em. Dettmann) Bharad. & Kumar 1972	Trilete, Cavate, sculptine, (finely patterned surface separated by a cavity)	Selaginellaceae : *Selaginellites* spp.
Playfordiaspora Maheshw. & Banerji 1975	Trilete, Cavate, sculptine (finely patterned surface separated by a cavity)	
Infraturma- *Laevigatomonoleti* Dybova & Jachowiez 1957		
Genera :		

Palynomorphs	Chief characters	Probable Affinities
Laevigatiosporites Ibrahim 1933	Monolete, Acavate, zonate, bean shaped, exine smooth	Polypodiaceae
Monolites Erdtm., 1947 Cooks, 1947, Chit. 1951 ex Pot. 1956	Monolete, Acavate, zonate, broadly oval to bean shaped smooth, thick exine	Polypodiaceae
Metamonoletes Singh & Kumar 1972	Monolete, broadly oval, Apiculate exine (coni)	? Marattiaceae
Dettmannites Singh & Kumar 1972	Monolete, subcircular or oval, Apiculate (Verrucae) distally differentially thickened exine.	? Marattiaceae
Turma - *Hilates* Dettmann, 1963		
Genera :		
Coptospora Dettmann 1963	Hilate, circular, exine thick, distally ruptured hilum.	Hepaticae, Sphaerocarpaceae, Ricciaceae, Riellaceae : *Geothallus* Spp.
Cooksonites (Pocock) Dettm 1963	Hilate, subcircular, Cingulate, distal hilum	Hepaticae
Aequitriradites Delc. & Sprum. em. Cookson & Dettm.	Hilate, zonate, subtriangular distal hillum, proximally tetrad mark. Zona membraneous, cavate near its inner margins.	Hepaticeae, Sphaerocarpaceae

Palynomorphs	Chief characters	Probable Affinities
Triporoletes (Rouseisporites) Mchedlishvili em. Playford 1971	Trilete, distal surface reticulate with muroid ridges, flask-shaped conical invagination in the equatorial radial regions.	Ricciaceae, clevaceae : *Riccia* spp.
Anteturma-*Pollentites* Potonie' 1931		
Subturma - *Monosaccites* Chitaley em. Pot, & Kremp 1954		
Infraturma - *Saccizonati* Bharadwaj 1957		
Genus :		
Callialasporites Dev em. Bharadwaj & Kumar 1972	Pollen grains oval to subglobose, monosaccate, equatorial saccus folded or frilled, smooth to granulose or microverrucose. Sometimes saccus becomes trinotched. Triradiate ridged or tetrad mark on proximal surface, Distal polar exine thinner.	Conifers - Podocarpaceae : *Apterocladus*, *Tsuga*
Subturma - *Disaccites* Cookson 1947		
Genera :		

Palynomorphs	Chief characters	Probable Affinities
Alisporites (Daugherty) Jansonius 1971	Pollen grains saccate, haploxylonoid, intrareticulate sacci, tenuitate	Coniferales
Falcisporites Leschik em. Klaus 1963	Pollen grains bisaccate haploxylonoid, intra reticulate saccite tenuita	Coniferales
Klausipollenites Jansonius 1962	Pollen grains bisaccate, haploxylonid, intra reticulate, distal thinner exine (tenuita)	Coniferales
Podocarpidites Cooks ex Couper 1953	Bisaccate, diploxylonoid, intrareticulate, renuitate	Coniferales- Podocarpaceae
Subturma - Polysaccate Cookson 1947		
Genera :		
Podosporites (Rao) Kumar 1984	Pollen grains tri to polysaccates, saccii intrareticulate, distally exine thinner at polar region.	Coniferales : Podocarpaceae
Turma -*Plicates* Naumova 1939		
Subturma - *Monocolpates Inversen* & Troel Smith 1950		
Genera :		

Palynomorphs	Chief characters	Probable Affinities
Cycadopites Monosulcites	Pollen elliptical, broader at equator than at pole tenuitate, end to end, exine smooth or scabrate	Cycadeoidales/Cycadales or Ginkagoales.
Turma-*Poroses* Naumova em. Potonie 1960		
Subturma- *Monoporines* Naumova 1939		
Genus :		
Classopollis (Pflug) Srivastava 1976	Spherical pollen, subequatorial circum-polar canal, a thickened equatorial band a distal cryptopore or operculum and proximal tetrad scar (vestigial)	*Pagiophyllum* spp. *Cheirolepis, Brachyphyllum* spp.
Turma - *Aletes* Ibrahim 1933		
Subturma - Azonaletes Luber emend Pot. & Kr. 1954		
Genus :		

Palynomorphs	Chief characters	Probable Affinities
Araucariacites Cookson ex.	Subcircular pollen, non saccate, inaperturate, exine ornamented and distally thinner at polar region (ulcus).	Araucariaceae : *Brachyphyllum mamillare*
Middle Gondwana Genera :		
Striatopodocarpites (Soritsch.) & Sed.) Bharad. 1962	Striated central body proximally disaccates pollen, distally tenuitate,	? Pteridospermae
Striatities (Pant) Bharadwaj 1962		
Distriatites Bharadwaj 1962		
Verticipollenites Bharadwaj 1962		
Lunatisporites Leschick 1955		

(ii) *Trilete, Cingulate-Auriculate spores*

This feature is recognised as thickenings in radial and equatorial regions of spore body produced by Matoniaceae, Dicksoniaceae, etc. (Table 18.2). The labrate margins also possess such thickenings (Harmomegathy). Such spores are *Matonisporites, Boseisporites, Lametatiletes* (Fig. 5) *Ischyosporites*, etc. (Table 18.2).

(iii) *Trilete Cingulate-Crassate*

Such features are observed in the spores of Gleicheniaceae, Lycopodiaceae e.g. *Gleicheniidites, Sestrosporites, Peregrinisporis, Dictyophyllidites*, etc. (Table 18.2).

(iv) *Trilete, cingulate-costate*

Cingulate-costate features indicated by thickenings at the equatorial region and costa (ribs or Muri) are developed on the distal surface of the spore body. Such morphographic features are present in *Contignisporites* (Figs. 6-7).

2. *Trilete, Costate-Murornate*

Costate (ribs or Muri) and reticulum or foveo-reticulum sculptures are developed on the distal or proximo-equatorial regions of the spore body. These are known in *Cicatricosisporites* (Figs. 2, 3), *Klukisporites, Ischyosporites* and *Lycopodiumsporites etc*. (Table 18.2).

3. *Trilete/hilate, zonate or flange bearing body exine*

The spores bearing equatorial extension of the outer layer (exoexine or zona/flange), inner layer (intexine) with or without cavity are developed in the sporomorphs, viz., *Aequitriradites* (Fig. 12), *Triporoletes* (Fig. 8), *Crybelosporites, Densoisporites* (Fig. 9) (perinate - Dettmann, 1963) etc. These spores are produced by the bryophytes (Hepaticopsida, and Selaginellid Plants) (Table 18.2). *Playfordiaspora* and *Lundbladispora* are trilete zonate/ flanged body exine.

4. *Trilete/Monolete, Sculptured or structured body exine*

Spores-pollen having sculptured exine on both the surfaces of the body or only on one face are known in *Osmundacidites, Concavissimisporites* (granulose exine) (Fig. 1), *Araucariacites, Callialasporites;* baculose/ verrucose exine in *Baculatisporites,*

Neoraistrickia, Sestrosporites, Metamonoletes, Dettmannites, conate exine *Ceratosporites, Coniatisporites;* cingulate-baculate-verrucate spores *Biformaesporites*' Foveo-reticulate spores *Klukisporites, Ischyposporites* (Fig. 4), *Trilites, Lycopodiumsporites,* etc. (Table 18.2). Among the structured spores bearing punctuate exine are *Boseisporites, Lametatriletes* and *Callispora,* etc.

All these spores are mostly the products of pteridophytic plants (Table 18.2).

5. *Apertural modifications*

Spores possess thick labrate margins in trilete forms, viz., *Matonisporites, Boseisporites, Lametatriletes,* etc., in monolete forms viz., *Monolites* (Fig. 11). Thickenings at interradial regions, i.e. in the vicinity of triletes mark, are present in *Dictyophyllidites*. Sculptured or structured labrate margins in trilete forms are present in *Trilites, Lametatriletes* (Fig. 5) and in monolete, *Metamonoletes* and *Dettmannites* (Table 18.2), Hilum (natural break down of exine) is known in *Aequitriradites* and *Coptospora* (Fig. 13) (distal hilum) and *Couperisporites* (proximal hilum), operculum is known in *Classopollis* (Fig. 22) also manifested with peculiar exinal organisation as presence of subequatorial circumpolar canal, a thick equatorial band, a distal cryptopore/operculum and a vestigial proximal tetrad scar), Table 18.2.

DISCUSSION AND CONCLUSIONS

Studies on living pollen grains revealed that various morphographic exinal features e.g. nature and organisation of sacci, exinal modifications, etc., that provide the mechanism to save the living inner genetical components and facilitate easy germination processes, etc., are very much sensitive to climate (Ueno, 1958, 1979, Guinet, 1987, Courtinat 1987, Hebda and Lott, 1973, Lamb, 1961, 1972, Kumar b in Press). The manifestation of cingulate-zonate/flanged, costate, Murornate structured or sculptured pattern of exine alongwith harmomagathic features as labrate margins of apertures, development of hilum or operculum and organisation of tenuitias/ulcus are considered very useful characters in deciphering the climatic conditions.

On the basis of exinal features it is apparent that Satpura basin had two types of land configuration, (1) one being the upland high altitudinal regions and (2) the others were low lying areas.

In upland regions gymnospermic conifers were thriving well and the low lying areas contained sufficient numbers of pteridophytes as well as some deeper low lying regions bryophytes were flourishing well. Such change in vegetational pattern have a close relationship with the environment of a particular region.

1. *Climate during Middle Gondwana*

(A) *Triassic climate* - Carbonaceous shales of Anhoni region belong to Denwa Formation (Nandi, 1991; and Vijaya, Pers. communication) contains prominance of haploxylonoid disaccates pollen (Table 18.1 Fig. 15). Ueno (1958, 1979) observed that species with haploxylonoid sacci (*Pinus pumila* and *Pinus koraiensis* grow in cool temperate and subpolar or subalpine zone.

Tiwari (1982), Tiwari and Tripathi (1987) opined that taeniae bearing pollen *Lunatisporites, Lueckisporites, Infernopollenites,* etc., (irregular ribbon shaped strips disposed on the proximal surface of the central body cap) had property to accumulate water for harmomegathy activity or for germinal exists which are related with seasonal mild fluctuations. Occurrence of zona/flange bearing spores further indicate that warmer/dry conditions with seasonal fluctuations were prevailing at low lying areas.

Climate during Upper Gondwana

(B) *Early Middle Jurassic climate :* At the Early Middle Jurassic time only one patch i.e. Parsapani belong to Jabalpur Formation of Satpura basin, was developed, in which dominance of diploxylonoid non-striated disaccate *Podocarpidites* (hollow sacci with infrabaculate/reticulate ornamentations) as well as haploxylonoid disaccates *Alisporites* as common member were present in upland areas. Ueno (1958, 1979) opined that species having diploxylonoid sacci (*Pinus thunbergia* and *Pinus densiflora)* flourish in temperate zone and relative abundance of both the forms is temperature controlled. Srivastava (1978) suggested warm, dry, temperate climate in neighbouring upland areas.

On the other hand at low lying areas where pteridophytic spores having mostly auriculate-punctate, crassate, sculptured/structured spores indicate warm with less humidity damp shaded areas with seasonal variations (Kumar b in Press, Ramanujam 1983, 1993).

(C) *Late Jurassic Climate :* During Upper Jurassic, palynoassemblages were recorded from different localities (2-II), had dominance of subsaccate frilled or vesiculate pollen *Callialasporites* followed by nonsaccate inaperturate (distally ulcate) *Araucariacites* showing upland flora. Courtinat (1987) performed an experiment on living pollen of *Aracucaria excelsa* R. Brown (belong to Araucariaceae) are identical to fossil pollen *Araucariacites* Cookson (1947). He observed that under warmer conditions such nonsaccate pollen are developed into subsaccate or zonate-cavate/flanged type of pollen which are very much identical to *Callialasporites monoalasporous* Dev (1961), *Callialasporites segmentatus* (Balme) Dev (1961) and *C. trilobatus* (Balme) Dev (1961). It reflects that during upper Jurassic the Satpura basin had developed some upland patches where warm, dry, temperate climate with seasonal variations might be prevailing.

Simultaneously in valleys of low lying areas plentiful numbers of pteridophytic plants were growing. The spores of such plants are characterised by cingulate-auriculate and cingulate costate, costate, muromate, sculptured/structured exine (Table 18.2). Tiwari(1982), Tiwari and Tripathi (1987), Kumar (b in press) envisaged that these features were related with the functions of water accumulation, harmonegathy activity employed during emergency germinal exists, seasonal mild fluctuations, as well as to save genetical components from the microbial activities or from incoming of adverse conditions, etc. Srivastava (1978), Ramanujam (1983, 1993); Kumar (1993, b, in press) suggested warm, humid, damp places with heavy rains and seasonal variations were prevailing at such places.

(D) *Early Cretaceous climate :* The Late Jurassic palynofloras were continued to Early Cretaceous time but the dominance of nonsaccate inaperturate *Araucariacites* is recorded in place of the dominance of the Late Jurassic subsaccate pollen *Calliaslasporites* (Bharadwaj *et al.*1972 and presently in Table 18.1), or vice versa also seen at Ellichpur area (Table 18.1; and Kumar, 1992). The subsaccate pollen with vesiculate nature of the radially elongated frilled exoexine might have developed from nonsaccate pollen under warmer dry climate conditions (Coutrinat 1987). This shows seasonal fluctuations in the climate in the upland areas. On the other hand, the occurrence of zonate/hilate or equatorially invaginated characters bearing bryophytic spores (Table 18.2) were proliferating at low

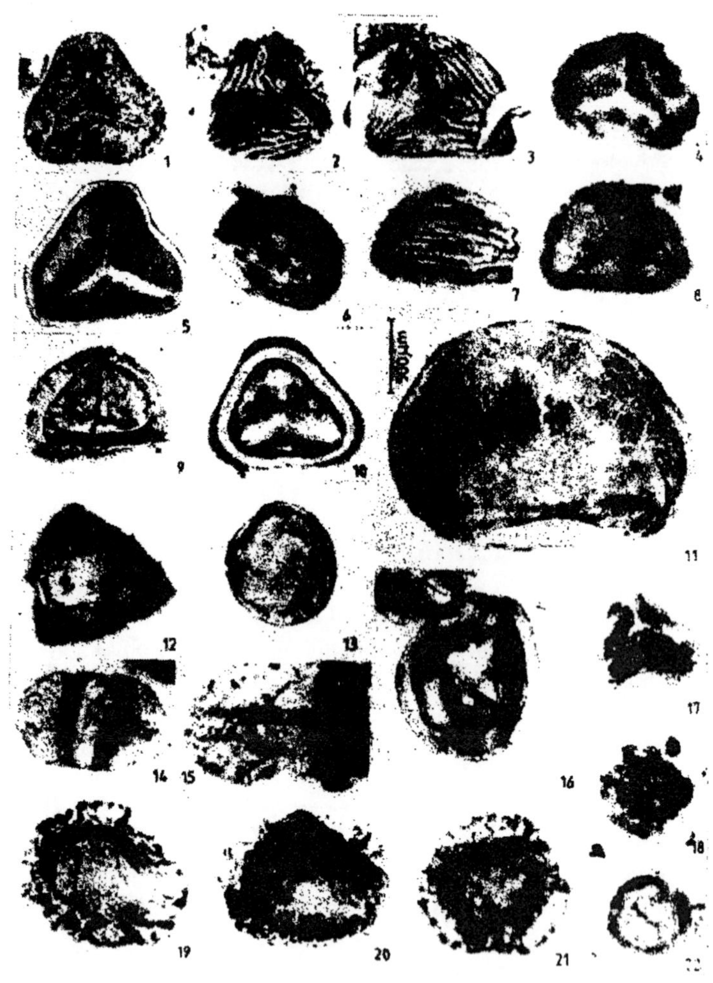

Plate 18.1

Plate 18.1 : All are magnified x 500; showing characteristic spore-pollen species occuring in the Upper Gondwana in Satpura basin. Scale is marked in the fig.11.

1. *Concavissimisporites* sp. Sl. No. B.S.I.P. 9993
2. *Cicatricosisporites australiensis* Sl. No. B.S.I.P. 9998
3. *C. ludbrooki*, Sl. No. B.S.I.P. 9998
4. *Ischyosporites haradensis*, Sl. No. B.S.I.P. 9995
5. *Lametatriletses indicus* Sl. No. B.S.I.P. 9996
6. *Contignisporites glebulentus* Sl. No. B.S.I.P. 9995
7. *C. dettmannii* Sl. No. B.S.I.P. 9992
8. *Triporoletes sehoraensis*, Sl. No. B.S.I.P. 9992
9. *Densoisporites mesozoicus* Sl. No. B.S.I.P. 9993
10. *Murospora Sp.* Sl. No. B.S.I.P. 9996
11. *Monolites indicus* Sl. No. B.S.I.P. 9991
12. *Aequitriradites triängulatus* Sl. No. B.S.I.P. 9998
13. *Coptospora kutchensis* Sl. No. B.S.I.P. 9990
14. *Podocarpidites ellipticus* Sl. No. B.S.I.P. 9992
15. *Klausipollenites australiensis* Sl. No. B.S.I.P. 9993
16. *Araucariacites ghuneriensis* Sl. No. B.S.I.P. 9998

17, 18. *Podosporites tripakshi* Sl. No. B.S.I.P. 9997, 9994

19. *Callialasporites dampieri* Sl. No. B.S.I.P. 9991
20. *C. segmentatus* Sl. No. B.S.I.P. 9990
21. *C. trilobatus* Sl. No. B.S.I.P. 9993
22. *Classopollis torosus* Sl. No. B.S.I.P. 9998.

lying areas which had warm, deeply shaded conditions with sufficient amount of humidity in the environment. Simultaneously the presence of costate, foveoreticulate or cingulate-sculptured/structured bearing exine mostly found in pteridophytic spores which are well developed under the same tropical-subtropical environment (Srivastava, 1978), Tiwari (1964). Kumar (b in press) presently described and studied several pterophytes (ferns) from the Pachmarhi hills and found that they are luxuriantly growing under such tropical environments.

From the perusal of the exinal features of spores-pollen occurring during Middle to Upper Gondwana sediments in Satpura basin can be also be envisaged or inferred the palaeoenvironment.

The following conclusions may be drawn :

1. The Satpura basin during the Triassic Period had continental conditions with warm, temperate, dry environment with restricted rains.

2. The Early-Middle Jurassic Period was also warm and dry with less humidity, but the Late Jurassic had some moderate climate with better humidity in low lying areas.

3. The Early Cretaceous Period had tropical to subtropical climate with good rains and some patches had swampy water logged conditions in low lying areas with seasonal variations, which are still prevailing at Pachmarhi hills in Satpura basin of Madhya Pradesh.

REFERENCES

1. Balme, B.E. 1957. Spores and pollen grains from the Mesozoic of Western Australia, C.S.I.R.O. Aust. Coal Res. Sect. T.C. : 1-48.

2. Bharadwaj, D.C. 1976. Palaeogeography of India during Gondwana Times and its bearing on the climate. Geophytology 6 : 2 : 153-161.

3. Bharadwaj, D.C. and Kumar, P. 1974. Palynostratigraphy of Parsapani coals, M.P. India. Geophytology 4 : 2 : 184-188.

4. Bharadwaj, D.C., Kumar, P. and Singh, H.P. 1972 Palynostratigraphy of coal deposites in Jabalpur Stage, Upper Gondwana India. Palaeobotanist 19 : 3 : 225-247.

5. Cookson, I.C. 1947. Plant microfossils from the lignites of Kerguelen Archipelago B.A.N.Z. Antarctic Res. Exp. 1929-31, Rep. A2 : 127-142.

6. Courtinat, B. 1987. Pollen grains from *Araucariacites* Cookson, 1947. Groups compared with the recent *(Araucaria excelsa* R. Brown). Rev. Micropalaeontologie 30 : 2 : 79-30.

7. Crookshank, H. 1936. Geology of the northern slopes of the Satpuras between the Morand and the sher rivers. Mem. geol. Surv. India. 66 : 2 : 242-272.

8. Dettmann, M.E. 1963. Upper Mesozoic Microfloras from south-eastern Australia. Proc. Roy, Soc. Victoria 77 : 1 : 1-148.

9. Dev. S. 1961. The fossil flora of the Jabalpur Series-3. Spores and pollen grains . Palaeobotanist 8 : 43-56.

10. Dutta, P.K. 1976. Climate during Upper Gondwana sedimentation in Peninsular India. Geophytology 6 : 2 : 170-173.

11. Feistmantel, O. 1876a. Note on the Gondwana age of some fossil floras in India. Rec. Geol. Surv. India. 9 : 2 : 27-62.

12. Feistmantel, O. 1876b. Note on the age of some fossil flora of India. Rec. Geol. Surv. Inst. 9 : 3 : 63-114.

13. Guinet, P. 1987. Geographic patterns of the main pollen characters in genus *Acacia* (Leguminosae), with particular reference to subgenus phyllodineae in : Blackmor, S& Ferguson, I.K. (eds) - Pollen and spores : form and function, Linnean Soc., London, : 297-311.

14. Gupta, K.M. 1966. Significance of the study of the cycadean fronds from the Upper Gondwanas of India (Rajmahal Hills). Symp. Florist Stratigr. Gondwld, B.S.I.I.P. Lucknow : 137 : 142.

15. Hebda, R.J. and Lott, J.N.A. 1973. Effects of different temperatures and humidities during growth on pollen morphology, an SEM study. *pollen spores* 15 : 563-571.

16. Kar, R.K. 1976 Microfloristic evidences for climatic vicissitudes in India during Gondwana. Geophytology 6 : 2 : 230-244.

17. Krishnan, M.S. 1982. Geology of India and Burma, C.B.S. Publishers & Distributors, India.

18. Kumar, P. 1973. The sporae dispersae of Jabalpur stage, Upper Gondwana, India. Palaeobotanist 20 : 1 : 91-126.

19. Kumar, P. 1986. Palynology of the Jabalpur Formation at Patbaba Ridge, Madhya Pradesh, India, I.S.G. Bull. 2 : 37-39.

20. Kumar, P. 1992 a. Palynology of Mesozoic carbonaceous sediments of Narsingpur district, Madhya Pradesh, India. Biol. Mem. 18 : 3 : 27-42.

21. Kumar, P. 1992b. Palynology of Mesozoic sediments exposed near Ellichpur, Maharashtra. Palaeobotanist 39 : 3 : 381-386.

22. Kumar. P. 1993. Palynodating of Chaugaon Bed exposed near Khatama caves, Hoshangabad district. Madhya Pradesh, India. Biol. Memories 19 : 1 : 1-16.

23. Kumar, P. a, in press. Palynodating of Mesozoic sediments exposed around Jatamao in Hoshangabad district, Madhya Pradesh. Biol. Memoires.

24. Kumar. P. b. in press. The present and the past pterophytes of the Satpura basin - A sporological Approach. Palaeobotanist.

25. Kumar, P.c. in press. Palynofossils from Pachmarhi Formation, Satpura Basin, Madhya Pradesh. Geophytology.

26. Kumar, P. and Kulshreshtha, S.K. 1979. Palynostratigraphical studies of carbonaceous shales from Kotri Narsinghpur district M.P. India. Geophytology. 9 : 1 : 52-61.

27. Lamb, H.H. 1961. Fundamental of climate in : Nairn A.E. (ed.) *Descriptive palaceoclimatology*. Interscience Publ. New York.

28. Lamb, H.H. 1972. Climate : present, past and future 1 : Mathuen & Co. Ltd. London.

29. Laskar, B. and Mitra, N.D. 1976. Palaeoclimatic vicissitudes in India during Lower Gondwana sedimentation. Geophytology 6 : 2 : 162-169.

30. Lele, K.M. 1962. Studies in the Indian Middle Gondwana flora -1. on *Dicroidium* from the South Rewa Gondwana Basin. Palaeobotanist 10 : 1, 2 : 42-62.

31. Lele. K.M. 1964. The problem of Middle Gondwana in India. Proc. 22nd internation. geol. Congr. New Delhi 1964. Sect. 9 (Gondwanas) : 181-202.

32. Lele, K.M. 1976. Palaeoclimatic implications of Gondwana floras. *Geophytology 6* : 2 : 207-229.

33. Maheshwari, H.K. and Kumar, P. 1979. A Jurassic mioflora from the Jabalpur Group exposed in Morand river near Morghat, Hoshangabad district, Madhya Pradesh. Geophytology 9 : 1 : 47-51.

34. Nandi, A. 1991. First record of Palynomorphs from Denwa Formation (Upper Gondwana) of pachmarhi area, Madhya Pradesh, Geol. Surv. India. Central Region, News 9 : 10.

35. Ramanujam, C.G.K. 1983. Floristics and Palaeoecological significance of the Lower Cretaceous microfloras of southern India : Proc. Symp. Cretaceous

India. Palaeoecol. Palaeogeogr. time boundaries, I.A.P. Lucknow 1982 : 89-93.

36. Ramanujam. C.G.K. 1993. Spore and pollen suites of some Upper Gondwana deposits of South India- A critical appraisal. Gondwana Geol. Magz. 1993. Spl. Vol. Birbal Sahni Cente. Nat. Symp. Gondwana India : 462-478.

37. Sastry, M.V.A. Aharya, S.K. Shah, S.C., Satsangi, P.P. Ghosh, S.C., Raha P.K. Singh, G. and Ghosh, R.N. 1977. Stratigraphic lexicon of Gondwana formations of India. Miscellaneous Publ. Geol. Surv. India. No. 36: 1-170.

38. Shah, S.C., 1976. Climate during Gondwana era in Peninsular India : Faunal evidences, Geophytology 6 : 2 : 186-206.

39. Shah, S.C. Singh, G. and Sastry, M.V.A. 1971. Biostratigraphic classification of Indian Gondwanas. Internat. Gond. Symp. Aligarh Muslim Univ. 1970. Ann. Geol. Dept. A.M.U. Aligarh 5 & 6 : 306-326.

40. Singh, H.P. 1966. Reappraisal of the mioflora from the Jabalpur Series of India with remark on the age of the beds, Palaeobotanist 15 : 1-2 : 87-92.

41. Singh, H.P. and Kumar, P. 1966. Some observations on the genus *contignisporites* Dettmann 1963. *Palaeobotanist* 15 : 1-2 : 93-97.

42. Singh H.P. and Kumar, P. 1972. Some new miospore genera from Upper Gondwana coals of India. Palaeobotanist 19 : 2 : 164-174.

43. Singh, I.B. 1976, Mineralogical evidences for climatic vicissitudes in India during Gondwana times. Geophytology 6 : 2 : 174-185.

44. Srivastava, S.K. 1978. Cretaceous spore-pollen floras : A Gobal Evaluation. Biol. Mem. 3 : 1 : 2-130.

45. Tiwari, R.S. 1982. Nature of straitions and taeniae in Gondwana saccate pollen. Geophytology 12 : 125-127.

46. Tiwari, R.S. and Tripathi, A. 1987. Palynological zones and their climatic inference in the coal bearing Gondwana of Peninsular India. Palaeobotanist 36 : 87-101.

47. Tiwari, S.D.N. 1964. Ferns of Madhya Pradesh. J. Indian Bot. Soc. 43 : 3 : 431-452.

48. Ueno, J. 1958. Some palynological observations of Pinaceae. Jl. Inst. Polytech. Osaka city Univ. 9 : 163-186.

49. Ueno, J. 1979. Pinus and Pollen analysis palaeolimnology Lake Biwa & Japanese Pleistocene. 7 : 348.

50. Vakhrameev, V.A. Jurassic floras of the USSR. Palaeobotanist 14 : 118-123.

51 Vredenburg, E.W. 1910. A summary of the geology of India, London.

ENVIRONMENTAL POLLUTION AND LAW

R.K. Sharma

A common saying of Mughal Emperor Jahangeer "Gar Firdaus Bar rue Jaminast", "Haminast Haminast Haminast" (if there is heaven on earth it is here only) depicts the natural beauty of pristine environment India had. Other poets and writers also praised the hurrying streams, orchards gleems, cool breeze of delight blowing in this land from Himalaya and Malayagiri. All this proves that India was rich not only in view of wealth but also in view of environmental purity and beauty.

Advances in science and technology have no doubt conferred many benefits on the society in the form of better and improved quality of good. The miracle of modern science and the towering achievements of technology have given us even a measure of mastery over nature[1]. But on the other hand, this advent of technology has also brought in its trail the problem of pollution of environment in which every creature lives. With the lapse of time and preferential treatment of industrial development, the problem of environmental pollution is becoming increasingly acute. It is more true in cases of developing countries like India.

Our forefathers were well aware of the importance of purity of air and water. Air, water and land are among the component parts of human body. As they knew that man himself was the main cause of pollution, in early times, so they took double measures to check or rather prevent mixing pollutants in the air and water. While they provided proper sanitation and drainage system in cities, there were rules of behaviour prescribed by the law givers for human beings to deal with such pollutants. The Prime Minister of Chandra Gupta Maurya, known as Kautilya, and Manu laid down such rules. Punishment was provided for violation of such rules. Chapter 36 of Book II of Kautilya's Arthsashtra provided fines for throwing dead bodies of cats, dogs, serpents, donkeys, camels and men inside the city[2]. Manu prohibited throwing urine, stool, speeting and throwing dirty material, blood or poisonous substance in water[3]. For leaving and throwing dirt of Royal roads, except in emergency, was punishable with fine of two 'Karkarshapana', and to pollute pure liquids was punishable with the penalty of 'Prathan Saha'[4].

On the same line Indian Penal Code, 1860 punished similar type of behaviour, as there was either no industrial pollution or if at all it was nominal, and the only cause of pollution at that time was human acts injurious to health etc. The relevant provisions are included in Chapter XIV dealing with offences afflicting the public health, safety, convenience etc. Section 269, 270, 273, 277 and 284 of the code punish acts likely to spread infection of diseases dangerous to life (negligent Sec. 269 and malignant Sec. 270) sale of noxious food or drink, fouling water of public spring or reservoir and negligent conduct with respect to poisonous substances respectively[5], with imprisonment upto 6 months (2 year in one case) or fine upto one thousand rupees or with both.

Though there were provisions in the Penal Code, 1860 and also in the Criminal Procedure Code, 1898 (section 133 dealing with removal of public nuisance), these were insufficient to control growing industrial pollution.

Just after two years of the United Nations Conference on Human Environment, an Act dealing with Water (Prevention and Control of Pollution) Act, 1974 was passed by the Parliament. The very purpose of the act was to provide for the prevention and control of water pollution and maintaining or restoring the wholesomeness of water, for establishment of boards, with a view to carrying out these purposes. Section 24 of the Act prohibited use of streams or well for disposal of polluting matters. Sections 25 and 26 created similar obligations. Sections 43 and 45 provided punishment for the acts done in violation of sections 24 and 25 at the first time and also subsequent violation. It provides a fair punishment[6].

Another Act, Air (Prevention and Control of Pollution) Act, 1981 was passed by the Parliament. Section 21 of the Air Act restricts use of industrial plants specified in the schedule to the Act without previous consent of the State Boards in an air pollution control area. Very wide powers are given to the Boards. Imprisonment or fine or both have been provided for various types of offences under the Act deoending on their gravity.

While these two specific statutes were inforce, the Parliament amended the Constitution of India in 1976 whereby a new Article (Article 48-A) was added in part IV of the constitution which included a policy directive for the States to endeavour to preserve, protect and improve the environment and to take legislative and other measures for the purpose[7]. Simultaneously, a new chapter dealing with Fundamental Duties was inserted which imposed a duty on the part of every Indian Citizen under Article 51-A (g) to protect and improve the natural environment including forests, lakes, rivers and wildlife and to have compassion for living creatures[7]. So far as these provisions are concerned, they are less helpful because of their non-enforceability in Court of Law. Only an independent, impartial and activist court can enforce these provisions under the umbrella of Article 21, "Right to life and Liberty", like the case *State of Himachal Pradesh V. Umedi Ram and others* (A.I.R. 1986 S.C. 847) wherein the court held that road in hill areas is life to a hillman, and therefore a pure environment is 'life' for every man[8]. It is to be noted that the High Court has taken the same line of arguments in *T. Damodar Rao V. Municipal Corporation, Hyderabad* (A.I.R. 1987 S.C. 171)[9]. However, looking into the infirmities of the existing law and an urgent need of a general law dealing with environmental pollution, to enforce the directive principle and to implement the decision taken in Stockholm conference, Parliament passed an exhaustive legislation the Environment (Protection) Act, 1986. Section 3 of the Act gives very wide powers to the Central government to take measures to protect and improve the environment. Sections 7, 8 and 9 impose duty on managers not to allow emission or discharge of environmental pollutants, to comply with procedural safeguards in handling hazardous substances and to furnish information to authorities. Section 15 prescribes penalty for contravention of the provisions of the Act and rules, orders and directions. The penalty is imprisonment for five years or fine upto one lakh rupees or with both. For further breach additional fine of Rs. 5000/- per day is prescribed. This Act applies to private as well as government industries, and the Heads of Departments shall be liable for violation according to section 17. Most outstanding feature of the Act is that it created liability of the officers directly incharge of the function of a company for their negligence. Section 16(2) created this liability for negligent or rash acts causing environmental pollution. "Criminalisation of corporate negligence needs to be seen as opening up a powerful approach that goes directly to the heart of the problem, which lies in the radical separation that has come about between the economic realm and the realm of values... once the economic realm gets detached from the realm of values and is no longer informed by a large sense of human purpose, it becomes a law into itself, and all other values are ruthlessly subordinated . . . "[10].

It is important to note that law cannot cure or purify the environment. Cure is again left to the scientists and technologists. What the law does is that it prescribes the norms of behaviour, duties to be performed by the management of industries. It also provides penal sanctions i.e., penalty for breach of these norms. Law prescribes measures of prevention to be taken compulsorily and punishment in case of failure. The existence of penal sanctions is a powerful restraining influence against selfseeking. Wayward or reckless behaviour. The treat of criminal sanctions alone has the power to compel corporate decisions makers to abandon the exclusively economic calculus of thought and action and for the first time to begin to base their behaviour on a serious consideration of the human consequences of their actions. In other words, it can force them to take society seriously and to begin to relate once again to the realm of values[11].

Thus we have number of statutes to control air, water and soil pollution. But mere letters of law are too weak to prevent or control pollution. had there been so, we would have not required Indian Penal Code or other penal laws. But it is not so. Efficiency of law depends on the efficiency and effectiveness of enforcement. It is sad state of law that we have strong laws but weak enforcement. To make them effective, their enforcement should not be avoided for one or the other pretext. It may not be out of place to quote Justice Krishna Iyer, "The politics of implementation is different from the menifestos of legislation and there is many a slip between the legislative cup and the law in action lip." If the laws are not properly implemented their existence is as good as their non-existence. Implementation of penal sanctions is more important in case violation of prescribed rules because the possibility of long years of imprisonment can ensure more wonderfully than anything else. That they subjected their own behaviour to the most searching and scrupulous scrutiny. Once the decision maker knows that his own neck is on the chopping block if he exposes others to unwanted risks, he has a powerful inducement to consider the totality of consequences of his actions and in that light to take the 'right' decisions.[12]

REFERENCES

1. Procreation of child in Test Tube is that wonder which confers the mastery to man over nature.

2. Section 26, throwing dirt on the road fine 1/8 pana; blocking road with muddy water 1/4 of a pana.

   Section 27, throwing dirt or muddy water of Royal High Way - double of what is provided in section 26.

   Section 30, throwing dead body of a cat, dog or serpent inside the city - fine three pana, dead body of a donkey, camel, horse or cattle - fine 6 pana, and for human corpus–fine 50 pana.

   The Kautilya's Arthasatra (2nd Ed.) R.P. Kanglay p. 94.

3. Manu Smriti 45, IV.

4. *Ibid.*, 187, IX.

5. Sections 269, 270, 273, 277 and 284 of I.P.C. 1860

6. Sections 43 and 45, Water (Prevention and Control of Pollution) Act, 1974.

7. Constitution (42nd Amendment) Act, 1976.

8. The High Court directed in this case, to the state government to get sanctioned budget and complete the roads as per plan.

9. High Court held that, "The enjoyment of life and its attainment and fulfilment guaranteed by Article 21 of the constitution embraces the protection and preservation of nature's gifts without which life cannot be enjoyed.. The slow poisoning by the polluted atmosphere caused by environmental pollution and spoilation should also be regarded as amounting to violation of Article 21 of the Constitution".

10. Economic and political weekly Vol. XXIV : 14 : p. 713 (1989).

11. *Ibid.*

12. *Ibid.*

SIGNIFICANCE AND IMPLEMENTATION OF WATER (PREVENTION AND CONTROL OF POLLUTION) ACT

S.K. Shringi

INTRODUCTION

Only 3% of the total global content of approximately 1.4 billion cubic kilometers of water, is fresh and suitable for human use. Of this again, about 77.2% is permanently frozen, 22.4% occur as ground water and soil moisture, 0.35% are contained in lakes and wetlands and less than 0.01% in rivers and streams (Water Resources, 1986). Thus freshwater is a very limited resource.

Agriculture, mainly irrigation, accounts for the largest single share of global water use. Conferences during the International Drinking Water and Sanitation Decade (1981-90) have stressed the importance of potable water in reducing diseases, the inter-dependence of soil and water and the multinational use of shared water resources. The demand for fresh water has increased and will increase with the rapid growth of population, agriculture and industry. As a result the supplies of fresh water are being depleted or polluted.

The requirement of clean water per person is about 2.7 litres per day or one cubic meter per year. Thus the minimal amount of drinking water at the global level needed annually is 5 billion cum or 5 cubic kilometers for survival. This requirement is not fulfilled in all parts of the world.

The annual rainfall as computed in India from 50 years data (1901-50) is about 105 cm. India receives 3 trillion m^3 of water, which is among the largest in the world. The country is rich in precipitation and surface water resources. Fourteen major river system share 83% of the drainage basin, accounts for 85% of the surface flow and serve 80% of the total population of the country. There are other 44 medium and 55 minor rivers which are mostly seasonal in nature.

Many of the major rivers also dry up during summer with no available flow for dilution of waste water discharged in to them. There is one more aspect and since independence, country has been harnessing water by dams and barrages for irrigation, power generation, for industries and drinking purposes and often there is repetitive use of water which only deteriorates the water quality. Often in the non-monsoon months there is hardly any water in a river system for dilution of waste water. There is now reasons of rethinking about such engineering works and attentions paid to quantitative aspects of water availability.

The subject of water management is thus a matter of urgent attention. Under the constitution, water, like land and forests, is a state subject and states have the legislative power to make and enforce legislations for surface and subsurface water. However, the union Government is empowered to promulgate legislation in areas covering interstate rivers and river valleys if such legislation is declared to be in public interest.

The water (Prevention and control of pollution) Act was passed in 1974 by the parliament. With in a short span of time almost all states of India have constituted the water pollution control Boards. The state of Maharashtra had passed its own pollution control Act, even before this Act of Parliament. The Rajasthan state had constituted the pollution control Board in the year 1976.

Pollution, according to the Act, means "such contamination of water or such alteration of the physical, chemical or biochemical properties of water or such discharge of any sewage or trade effluent or of any other liquid, gaseous or solid substance in to water (whether directly or indirectly) as may, or is likely to, create a nuisance or render such water harmful or injurious to public health or safety, or to domestic, commercial, industrial, agricultural or other legitimate uses or to the life and health of animals or plants or of aquatic organism".

WATER RESOURCES

An estimate in 1974 (Nag B.S. and Kathpalia G.N. 1975, water Resources in India) showed that in India, out of the annual precipitation of 400 million hectare metre (m ham), the water utilization was only 9.5% which was estimated to rise to a maximum of 26% in 2025 AD.

For the 1900 M m³ of water available/year the projected water use pattern for India (2000 AD) is given in Table 20.1. It is clear from the table that irrigation (including for live stock) accounts of largest share 79.6% followed by power 13.7%, domestic 3.5% and industrial 3.2% uses.

Table 20.1- Water Use (India) 2000 A.D. (Available water 1900 million cubic metre per year)

	In 1000 million cubic metre per year		
Uses	Taken	Consumed	Returned
Irrigation and Live Stock	869	783	86
Power	150	5	145
Industry	35	10	25
Domestic	38	8	30
Total	1092	806	286

Source : Chaudhuri, 1982.

One of the obvious hazards of environmentally unsound irrigation projects is water logging which brings in salinity. Nearly 50% of the water logged land is the result of surface flooding. Water logged area of west Bengal, Punjab and Haryana together accounts around 60% of water logged areas of country.

Most of the urban communities have the drinking water supply. Out of 3121 towns, the 2092 have an organised water supply. The coverage by number of towns and by population is given in Table 20.2. At present there is a low daily per Capita supply, in efficient distribution, high distribution leakage and ill managed systems. Progress of rural water supply is very poor because of the fact that there are large number of villages in the country.

Since last three decades, the urban communities have been supplied with drinking water, but the provision of collection, treatment and disposal of domestic waste water has been lagging behind, resulting in pollution of our water bodies. Most of our rivers, estuaries, streams are badly polluted. An action plan for control of this waste water has been prepared by the

Central Board for Prevention and Control of Pollution. The report shows that in class I cities, with a pollution of more than one lac, piped water has been provided. Some of these have been able to provide proper modern treatment plants. However, it has not been possible to provide sewerage system to carry and dispose off waste water to about 57% of this. Even large metropolitan cities have only partial sewerage system. The biological pollution is therefore quite high resulting in water borne diseases.

Table 20.2 : Water Supply coverage in Towns (1978)

Class of Towns	Population range '000	Coverage by water supply	
		By Number (ratio)	By 1971 population (percent)
I	100 or more	149/150	95.2
II	50 to 100	206/221	84.7
III	20 to 50	542/652	76.0
IV	10 to 20	649/988	60.8
V	5 to 10	423/821	57.2
VI	5 or less	123/289	40.1
		2092/3121	20.4/109=82.9

(Source : Chaudhuri, 1982) Chaudhuri, N. 1982. *Water and Air Quality Control : The Indian Context.* The Tenth Anniversary of the 1972 stock home conference on Human Environment, Central Board for the Prevention and Control of Water Pollution, New Delhi.

SOURCES OF WATER POLLUTION

Most of our water bodies including rivers, streams and lakes are highly polluted. Pollution of water bodies are mainly due industrial effluents, agricultural waste and domestic sewerage. Both from domestic as well as from organic chemical industries, septic conditions have been established and there are reported fish kills and destructions of aquatic life. These waters are unsafe for drinking and other purposes.

Table 20.1 indicates that quantity of water returning after use, which actually becomes waste water is the highest in thermal power, followed by irrigation run-off. The former is indeed a cause of concern particularly when it is coupled with air pollution and solid wastes.

A surveys of the industrial cities of India has shown that industrial waste water constitutes by volume between 8 to 16% of the total waste water generated, and the remaining 84-92% is form domestic sources. With rapid pace of industrialization, by 2000 AD these figures are expected to reach about 33% and 67% respectively. In terms of BOD (Biochemical Oxygen Demand) at present the pollution load is 50% each for industrial and domestic waste water.

A study of the Yamuna, the main source of drinking water for Delhi, by the central board of preservation and Control of Water Pollution, showed that the waste water flowing into a river carries 3,00,000 cubic metres of the sewage and 20,000 cubic metres of toxic industrial waste every day. The various industrial units in Delhi producing textiles, chemicals soap, rubber, engineering goods are said to be producing liquid waste, for which there is no proper control measures.

Analysis of our drinking water supplies in cities can positively unearth, not only chlorine compounds but other poisonous chemicals, too. Some time back, a report of the National Committee on Environmental Planning and Co-ordination (India) indicated that sea fish from off the Gujarat Coasts are contained with mercury.

Studies by Bhabha Atomic Research centre at Trombay show that chloro-alkali plants in Bombay area were releasing 0.5 to 1 pound of mercury for every ton of chlorine produced. About 97 percent of the released appeared in waste water and sludges which were discharged into sea, causing pollution of coastal waters. Such pollution leads to biological concentration of the mercury in its most toxic form in aquatic organism including fish.

Over and above these chemicals, pathogenic micro organism also abound in many of the drinking waters. Fertilizers and pesticides also find their way into the ground waters. Similarly industrial effluents, when they are dumped on land, sink in, contaminating the ground water. A large majority of river surveys have established this. Due to industrial effluents with high BOD and toxic wastes such as heavy metals, cyanides, arsenic phenols or non-biodegradable wastes such as pesticides etc. the water is unfit even for agricultural and irrigation purposes.

WATER BORNE DISEASES

Water pollution causes serious water borne diseases. Most of our cities that have piped water supplies are able to supply water intermittently. The distribution system does not remain under pressure and if the sewers are laid in the vicinity of water pipes, it can suck in ground air and there have been many reported cases of contamination of supplies. There has been many cases of water borne diseases, due to this contamination and several epidemics of viral hepaths have been reported from the urban areas.

Water borne diseases often assumes epidemic proportions in India. Cholera, Typhoid, Gastero-entertis and hepatitis are all caused by contaminated waters. Bombay and Delhi are often visited by hepatitis, when the drinking water gets contaminated by sewage. Guineaworm, Tape worm, Round worms and their eggs are water borne. Mosquitoes that spread Malaria, filaria and Dengne breed in stagnant waters and transmit these ailments.

POLLUTION CONTROL BOARDS

It is only after the passing of water Act in 1974, the water pollution work started. The Central Board for the prevention and control of water Pollution conducted a survey on waste water collection, treatment and disposal in class I (over 1 lac people) and class II (50,000 to 1 lac people) and published the results in the forms of two very valuable reports CUPS/4/1978-79 and CUPS/61979-80 respectively first time.

State Water Pollution Control Boards have to take comprehensive programmes for the prevention, control or abatement of pollution. Their first work was to prepare inventories of industries in their jurisdiction, in the whole state.

For assessing the damage, the pollution Control Boards have to take comprehensive surveys of the rivers in these areas. Some of the perennial rivers are being surveyed as a part of world health organisation. Global Environmental Monitoring Programme. Some Boards have taken up such surveys with the specific purpose or advising the Industries and towns about the quality of water at different places in the streams, as also the tidal portions of the stream where the effluents could be safely discharged with minimum of treatment. These surveys assist the boards in classification of streams to be able to decide which part of the river is good for drinking water withdrawal, recreational, industrial and irrigational uses.

The ISI standard for the disposal of effluents are different for different situations and are shown in Table 20.3. Four standards have been determined by the ISI for the guidance of Water Pollution Control Boards. These standards are for discharge into (a) a town sewer (b) a stream from where water is used for domestic purpose or for industries, irrigation, recreation etc. (c) the marine environment, which may be either as estuary or the sea, and (d) the land for treatment or irrigation. These guidelines have been tentatively accepted, as "Standards", though in some Boards, the standards are issued according to local situations and conditions, of streams and may be slightly higher or lower than the guidelines prescribed by ISI. In all the causes, standards are for the specific time and needs periodic revisions.

Many of the Boards have setup their own laboratories, where analytic work is carried out. In addition some private laboratories and laboratories of training institutions have also been recognised to supplement the analytic work.

Some boards have also taken up training programmes for their own staff and for the technical personnel working in other industries, municipalities and in Government sectors. In this activity, the Central Board is also actively co-operating with state Boards, Refresher courses and workshops are also being organised for treatment of industrial effluents and for the maintenance of treatment plants, so as to enhance their knowledge and experience and to educate them in better management. Treatment plants that are being installed or under construction are managed by the boards personnels.

It is very necessary to expand the activity as more experience is being gained and as more knowledge is available for the treatment processes, that have been successful in our country. Our meteorology. Financial limitations, Research and Development also forms an important parts of the board activity.

The Boards have to give "Consents" to the industries to create new discharges into water courses according to specific standards. For this purpose also, the quality of effluents have to be ascertained by scientific analysis and by study of a number of parameters. Monitoring of effluents

Table 20.3 : ISI standards for the disposal of industrial effluents

Sr. No.	Parameters	Tolerance limits for Industrial effluents discharged			
		IS : 3306-1974 Into Public sewer	IS : 3307-1977 First Revision On land for Irrigation	IS : 2490 (Part 1) 1974 Into Inland Surface waters	IS : 7968-1976 Marine disposal
1	*2*	*3*	*4*	*5*	*6*
1.	pH	5.5 to 9.0	5.5 to 9.0	5.5 to 9.0	5.5 to 9.0
2.	Temperature (Max.)	45°C	-	Shall not exceed 40°C in any section of the stream within 15 meters down stream from the effluent outlet.	40°C at the point of discharge.
3.	Total Suspended Solids mg/1	600 Relaxable to 750 by the local authority	-	100	(a) For process (b) For cooling water effluent Total suspend-ed matter of influent cool-

1	2	3	4	5	6
					ing water plus 10 per cent.
4.	Particle size of Total Suspended Solids.	-	-	Shell pass 850 Micnons IS : SIEVE (See IS : 460-1962)	(a) Flotable Solids : Max. 3mm. (b) Settable Solids : Max. 850 Microns.
5.	Total Dissolved Solids mg/1	2100* (Inorganic).	2100	-	-
6.	Biochemical Oxygen Demand for 5 days at 29°C mg/1 (Max.)	500 subject to relaxation or tightening by the local authority.	500	30	100
7.	Chemical Oxygen Demand mg/1	-	-	-	250
8.	Oil and Grease mg/1	100	10	10	20
9.	Chloride (as C1) mg/1	600*	600	-	-

1	2	3	4	5	6
10.	Phenolic Compounds mg/1	5 (as C_6H_5OH) Relaxable to 50 by the local authority when secondary treatment of sewerage is carried out	-	1.0	5 (as C_6H_5OH)
11.	Cyanides (as Cn) mg/1	2.0	1	0.2	0.2
12.	Sulphate (as SO_4)mg/1	1000*	1000	-	
13.	Sulphides mg/1	-	-	2.0	5.0
14.	Insecticides mg/1	-	-	absent	-
15.	Pesticides mg/1	-	-	-	(a) Organophosphorous Compounds (as P) ..1 (b) Chlorinated hydrocarbons (ascl) ..0.2
16.	Total Residual Chlorine	-	-	1	1
17.	Fluoride (as F) mg/1	-	-	2.0	15
18.	Boron (as B) mg/1	2*	2	-	-

1	2	3	4	5	6
19.	Arsenic (as As) mg/1	-	-	0.2	0.2
20.	Percent Sodium	60*	60	-	-
21.	Cadmium (as Cd) mg/1	-	-	2.0	2.0
22.	Copper (as Cu) mg/1	3	-	3.0	3.0
23.	Lead (as Pb) mg/1	1	-	0.1	1.0
24.	Hexavalent Chromium (as Cr) mg/1	2.0	-	0.1	1.0
25.	Mercury (as Hg) mg/1	-	-	0.01	0.01
26.	Nickel (as Ni) mn/1	2	-	3.0	5.0
27.	Selenium (as Se) mg/1	-	-	0.05	0.05
28.	Radioactive Materials				
	(a) Alpha Emitters uc/m1	10_{-7}	10_{-9}	10_{-7}	10_{-8}
	(b) Beta Emitters uc/m1	20_{-6}	10_{-8}	10_{-6}	10_{-7}
29.	Ammonical Nitrogen (as N) mg/1	50	0	50	50
30.	Zinc (as Zn) mg/1	15	-	5	5.0

These requirements shall apply when after treatment, the sewage effluent is disposed off for irrigation and on land.

of the industries is one of the important activities of the boards. This keeps a vigil, whether the treatment plants installed are running efficiently and that pollution is being controlled.

For the design of the industrial effluent treatment plants the state boards gives general advice and the guidance to the industries. But the detail designs and responsibility for the construction of treatment plant lies with the industries, again industries can go for consultations with Boards.

There is a great variety of industries established in the last three decades in India. Some of them produce effluents that are non-polluting or easy to be treated. There are many other effluents that can be easily bio-degradable and could be treated by low cost treatment plants. There are some industries, which discharges effluents that are not easily bio-degradable and are very toxic in nature, these are difficult to be treated. The following list of industries that have been established in our country and which needs appropriate and suitable effluent treatment:

1. Oil Refineries
2. Petrochemical industries
3. Fertilizer industries
4. Cellulose industries
5. Paper and pulp industries
6. Starch industries
7. Sugar industries and distilleries
8. Dye Stuff industries
9. Textile and Rayon industries
10. Pharmaceutical industries
11. Pesticides industries
12. Tanneries and leather industries
13. Steel Mills and Mining
14. Chemical industries
15. Chlor-alkali industries
16. Cement industries.

THE WATER (PREVENTION AND CONTROL OF POLLUTION) CESS ACT

This Act was also passed by parliament in the year 1977. As per this Act, anyone consuming water has to pay certain amount of cess depending on (1) Whether the industry is using the water for industrial cooling, spraying in mine pits or boiler feed (2) for domestic purposes (3) in processing where by water gets polluted and pollutants are easily bio-degradable, and (4) in processing whereby water gets polluted and the pollutants are not easily bio-degradable and are toxic.

Those industries that had installed a suitable treatment plant for the treatment of industrial effluents can get a rebate of 70 percent of the cess payable.

DISCUSSIONS

The boards have been able to collect a large quantity of information about the quantity of water used by individual industries and the quality of waste water. Some industrial effluents could be easily treated by different methods. The biological treatment has been sophisticated like activated sludge or Extended Aeration. It could also be possible to treat effluents by anaerobic or aerboic lagoons, Oxidation ponds, Several Industries are now attempting to minimise the quantity of waste effluents by better house-keeping and segregation of effluents and by recycle.and reuse of wastes. As a result of research and experimentation, many new treatment systems have been devised. The treated effluents are being used on land drained into irrigation canal and many chemicals and metals are being recycled. The industries are thus able to recover substantial money towards cost of treatment, besides solving the effluents treatment problems.

The boards have realised that some industries may pose large funding problems. In such conditions, industries are advised to chalk out these treatment programmes, and complete the project within some stipulated time. For small industries, after treatment, the board allows the effluents for solar evaporation. But this thing has to keep in mind that no toxic effluent could make their way in to underground water.

Studies are being undertaken on land treatment since in a tropical country, it is advisable to use effluents for irrigation and for growing crops of different types and for reforestation. There are some industries that have not yet installed the treatment plants as per some reasons are now persuaded to work out time-bound treatment programmes.

With greater awakening among our people, the social and legal obligations and availability of knowhow from other countries and our indigenous experiences and researches, a stage of take-off has been reached. With greater concern in the Government and in the Community for a cleaner and better environment, we would be able to force the industries to install appropriate effluent treatment plants and keep our environment clean.

THE ENVIRONMENT (PROTECTION) ACT, 1986 – AN ASSESSMENT

Dr. Paramjit S. Jaswal and Dr. Nishtha Jaswal

Environment planning and preservation is today the concern of all. With the change of concept of social welfare state, now the state is not merely concerned with the maintenance of law and order in the society but has also assumed all the duties which touch public life, public welfare and hence it becomes an essential duty of the state to provide every citizen a healthy environment to live in.[1] However, different dimensions of the problem of environment protection and its management have taken a serious turn in the present era. If we are serious about the quality of life not only for the present generation but also for the future generation, then we have to maintain ecological balance, protect the environment from pollution and manage and improve it properly.[2]

The world community's resolve to protect and enhance the environmental quality found expression in the decisions taken at the United Nation's conference on Human Environment held in Stockholm in June 1972. The Government of India also participated in the conference and strongly voiced the environmental concerns. While several measures had been taken for environmental protection, both before and after the conference, the need for general legislation further to implement the decision of the conference had become increasibly evident.[3]

In India, although there were existing laws dealing with directly or indirectly with several matters, it was necessary to have general legislation for environment protection. Existing laws were generally focused on specific types of pollution or on specific categories of hazardous substances.[4] Some of the major areas of environmental hazards were not covered. Also the law is being rapidly out distanced by the accelerating pace and expanding scale of impacts on the ecological basis of development.[5] There were also inadequate linkage in handling matters of industrial and environmental safety. Control mechanisms to guard against slow, insidious build up of hazardous substances, especially new chemicals, in the government were weak.

In view of what has been stated above, the general legislation on environment, that is, the Environment (Protection) Act, 1986 was enacted.[6]

The Act extends to the whole of India[7] Section 2(a) of the Act defines environment as under :

> "environment" includes water, air and land and the inter-relationship which exists among and between water, air and land, and human beings, other living creatures, plants, micro-organism and property.

Thus, this definition is inclusive and not exhaustive and it dynamically recognises the inter-relationship which exists between water, air, land and human beings, other living creatures, plants, micro-organisms and properly. Section 2 also defines various other terms like "environmental pollutant"[8] "environmental pollution"[9] and "hazardous substance"[10] etc. All these terms or definitions are 'exhaustive' and not 'inclusive'. An inclusive definition has the distinct advantage for the exercise of vast rule making powers under the Act and for a more effective enforcement of the Act. On the other hand, exhaustive definitions, in an evolving field like environmental control, are likely to head to recourse to judicial interpretation of highly complex scientific and technological matters, whose complexion is ever changing as knowledge accumulates dynamically. Therefore, it is, suggested that other definitions should also be inclusive and not exhaustive.[11]

Under section 3 of the Act, power has been given to the Central Government to take all such measures as it deems necessary or expedient for the purpose of protecting and improving the quality of the environment and preventing, controlling and abating environmental pollution.[12] It is also contemplated that the central government may constitute an authority or authorities by such name or names as may be specified, for the administration of the Act.[13] The meeting of the experts[14], recommended the establishment of a National Environmental Protection Authority (NEPA).[15] The Meeting recommended that NEPA should be participatory authority and its structure must be such that it should accommodate diverse channels of information concerning environmental pollution and protection.[16] Such a participatory structure is commended by basic principles of democratic governance as well. It was further recommended that the NEPA be constituted by a collegian comprising about twenty people which should include eminent public citizens, scientists, social activities in the field of environment, representatives from the labour, industry, consumer and women's groups.

It was also suggested that National Environment Protection Authority would set up Environment Impact Assessment Group (EIAG). It would consist of eminent environmentalists and scientists with interest in public interest advocacy. Its main purpose would be (a) to keep on record and disseminate information in the field of ecology (b) to provide professional consultancy services to help investigators to get technology assessment and project evaluation done and to help citizens and pubic interest group. (c) to advance people's knowledge in the field of environmental law. (d) to undertake immediate techno-ecological analysis of ecological and environmental disasters.[17] The meeting of experts also suggested the setting up of Standard and Enforcement Division. This division of the Authority would be an independent division of NEPA and will be responsible for setting standards associated with the protection of the environment, like emission standards, noise standards etc.[18]

It is submitted that the above suggestions of the meeting of experts to set up above authorities are very good and they will cover almost all the aspects of the administration of the Environment (Protection) Act, 1986. But it is not clear what relationship these bodies or authorities will have with the State Boards. It is clear, on the other hand, that no lessons were learnt from the failure of the Boards before enacting the new Act. The new Environment Protection Act seems to be oblivious of the previous Pollution Control Acts and the Administrative Bodies set up under them.[19]

One of the mot important recommendations of the meeting of the experts is of setting up of "environment Courts."[20] It was observed that an effective and easily accessible system of handling complaints, grievances and disputes relating to the environment is essential to the full exploitation of the potential of the Environment Protection Act, as well as all related legislations on environment. This suggestion is in consonance with the suggestion which the Supreme Court of India gave in *M.C. Mehta* v *Union of India*[21], The Supreme Court suggested to the Government of India that since cases involving issues of environmental pollution, ecological destruction and conflicts over natural resources are increasingly coming up for adjudication and these cases involve assessment and evaluation of scientific and technical data, it might be desirable to set-up "Environmental Courts" on the regional basis with one professional judge and two experts drawn from the Ecological Science Research Group keeping in view the nature of the case and the expertise required for its adjudication.[22]

It is submitted that this is a very healthy suggestion and Central Government must implement it at the earliest.

Section 4 gives the Central Government a power to appoint officers with such designation as it thinks fit for the purpose of the Act. Section 5 of the Act gives power to the Central Government to give directions which includes the power to direct:

(a) the closure, prohibition or regulation of any industry, operation or process; or

(b) stoppage or regulation of supply of electricity or water or any other service.

It is submitted that this provision provides flexibility and range of deterrent and preventive mechanism to the governmental authorities for achieving the objective of the Act.[23]

Persons carrying on industry, operation etc. are not allowed the omission or discharge of environmental pollutant in excess of the standards.[24] Persons handling hazardous substances are also required to comply with the procedural safeguards.[25] Where the discharge of any environmental pollutant in excess of the prescribed standards occurs or is apprehended to occur due to any accident or any other unforeseen act or event, the person responsible for such discharge and the person in charge of the place where such discharge occurs shall be bound to prevent or mitigate the environmental pollution caused by such discharge.[26] Such occurrence of discharge is also required to be informed to the authorities concerned and remedial measures should be taken to prevent or mitigate the environmental pollution.[27] Whatever expenses are incurred in respect of such measures, shall be recovered from the person concerned as arrears of land revenue.[28]

It is submitted that this provision will make the persons in charge of hazardous substances or pollutants, more responsible.

Provisions have also been made in the Act regarding the powers of entry and inspection[29] of the industrial plant; to take sample and procedure to be followed therein;[30] to set up environmental laboratories[31] and to appoint government analysts.[32]

Section 15 of the Environment Protection Act, 1986 is one of the most important provisions which deals with penalty for conservation of the provisions of the Act and the rules, orders and directions thereunder. Each failure of compliance or contravention is punishable with a term of imprisonment which may extend to five years or with fine which may extend to one lakh rupees or with both . For each act of failure to comply or contravention, happening after the conviction for such failure or contravention, an additional fine of rupees five thousands per day is prescribed.[33] And if such failure or contravention continues beyond a period of one year after the date of conviction, the offender is liable to imprisonment for a term which may extend to seven years.[34]

It is submitted that the punishment are clearly intended to be harsh but there are some lacunae in this provision. First, no minimum punishment is prescribed. All the punishments mentioned section 15 use the words "may extend to" and prescribed the maximum limits only. This gives wide discretion to judge and courts in awarding the sentence. It is suggested that the section should prescribe both minimum and maximum limits of the punishment. Otherwise, the wide discretion of the judges in awarding the sentence might weaken the intended deterrent impact of the Environment Protection Act. Secondly, the liability for the punishment for continuing offences arises only "after the conviction" for the first such failure or contravention. But since the conviction can be appealed against, in some cases right up to Supreme Court, this process may consume number of years, hence the basic purpose of awarding the second sentence is defeated. Thirdly, when violation and contravention continuing beyond one year period after conviction is punishable with enhanced term of imprisonment, there is no provision for a minimum mandatory punishment.[35] It is suggested that if person has committed an offence by violating the environmental law and after the trial for that has begun if a person commits another offence, even before the conviction for the first offence, then for the second offence an enhanced punishment should be awarded. This will definitely have some different effect as is intended by the Environment Protection Act, 1986.

Section 16 adequately pierces the corporate veil. It provides that where the offence is committed by a company then every person who was directly in-charge of the business of company, at the time when the offence was committed as well as the company shall be deemed to be guilty of an offence an shall be proceeded against and punished accordingly.[36] Whenever it is proved that the offence was committed with the consent or connivance of any director, manager, secretary or other officer of the company then such person shall be deemed to be guilty of an offence and shall be punished accordingly.[37] However, if it is proved that such officer exercised all due diligence to prevent the commission of the offence or it was committed without the knowledge of such person no liability of such officer shall be there.[38]

Section 17 is an innovative provision which provides that where the offence is committed by any government department or with the connivance of the head of the such government department, then such head of the department shall be deemed to be guilty of offence and shall be liable to be prosecuted against and punished accordingly. However, if it is proved that the offence was committed without the knowledge of such head or he exercised due diligence to prevent the commission of such offence then no liability of such heads of the department shall be there. The existence of such a provision will make head of the departments more careful and alert in dealing with environmental matters. When they know that they can be held liable personally, they would obviously play their role with utmost caution. This in turn should be able to bring in positive results in matters relating to environment. This provision is self sufficient in as much as it protects such officers who have acted with due diligence. In spite of this if still the offence is committed, they would not be guilty.

It is submitted that all the good intentions of legislature of providing harsh punishment are nullified by section 24 (2) the Act which provides that where any act or omission constitutes an offence punishable under this Act and also under any other Act then the offender found guilty of such offence shall be liable to be punished under the other Act and not under this Act. This provision is anomalous, since many offences would also be punishable under the provisions of previous pollution laws which prescribed a lesser punishment and hence in such cases the new Act will only prove to be paper tiger.[39]

Recently, it has been realised that for the proper implementation of any legislation, the public participation is a must. It is this realisation, belated but crucial, which has generated the trend in the progressive legislations notably since 1985 which allocate a significant role in the implementation and enforcement of laws to concerned non-government organisations and social action groups.[40] However, unlike other progressive legislations, the Environment Protection Act, 1986 does not recognise expressly the role of public participation in enforcement or implementation. Section 19(b) places a restriction on the powers of courts to take cognizance of any offence under the Act unless a person complaining of an alleged offence under the Act has given a notice of sixty days, of his intention to make complaint, to the Central Government or a designated authority or officer. This provisions militates comprehensively against public participation in the enforcement of the salutary provisions of the Environment Protection Act, 1986. It is submitted that this limit to sixty days should be removed and when only the public participation can become useful for the enforcement of the Act.[41]

The Act, however, clearly reflects the profound anxiety of the law makers to give effect to the solemn resolutions of the Stockholm Conference of 1972 on human environment.

REFERENCE

1. P.S. Jaswal. "An Appraisal of legislation and Judicial Role in Environment Protection", in R.C. Delela, Shashi Kant and Shama Vohra, *Environment and Pesticide Toxicity,* 185-189. (Proceedings of 8th Annual session and Symposium of the Academy of Environmental Biology, India))

2. P.S. Jaswal and Nishtha Jaswal, "Environment Protection and its management: A Study of Legislative Action and Judicial Activism in India "In R.K. Sapru and Shyama Bhardwaj, *The New Environmental Age,* (1990)

3. See the "Statement of Objects and Reasons" of *The Environment (Protection) Act,* 1986.

4. For example, under the Indian Penal code, section 268 deals with "public nuisance" under which noise pollution or other kind of pollution which creates nuisance can be controlled. Sections 269-271 deal with the negligent act spreading infection of disease. Sections 272-276 deal with adulteration of food. Section 277 deals with water pollution in certain cases. Section 278 deals with pollution of atmosphere obnoxious to health. Sections 284-286 deal with negligent handling of poisonous and explosive substances. Sections 133-143 of the Criminal Procedure Code also control the "Public nuisance". Some of the specific legislations passed after the Stockholm

Conference, 1972 were: The Wild Life (Protection) Act, 1972; The Water (Prevention and Control of Pollution) Act, 1974; The Air (Prevention and Control of Pollution) Act, 1980; The Forest Conservation Act, 1980; The Bhopal Gas Leak Disaster (Processing of claims) Act, 1985; and Motor Vehicle Act, 1988.

5. See *Our Common Future- The World Commission on Environment and Development* 21 (1987)

6. Act No. 29 of 1986 published in the *Gazette of India, Extraordinary*, Pt. II Section 1, dated 26 May 1986. This Act came into force, *w.e.f.* 19 November 1986.

7. Section 1

8. Section 2(b): "Environmental Pollutant" means solid, liquid or gaseous substance present in such concentration as may be, or tend to be injurious to environment.

9. Section 2(c): "Environmental pollution" means the presence in the environment of any environmental pollutant.

10. Section 29 (e): "Hazardous substance" means any substance or preparation which, by reasons of its chemical or physio-chemical properties or handling is liable to cause harm to human beings, other living creatures, plants, micro organism, property or the environment.

11. This can be done if the word "means" is substituted by the word "includes" in these definitions. See Upendra Baxi, *Environment Protection Act: Agenda for Implementation* 6 (1987).

12. See section 3 (1) and 3(2) of the Act.

13. Section 3(3) of the Act.

14. A meeting of Experts to consider the ways in which the Environmental Protection Act, 1986 may be effectively implemented was convened by the Consumer Education and Research Centre and Indian Law Institute on 22-24 August, 1986. See *supra* note 11 at 1.

15. *Supra* note 11 at 8

16. *Ibid.*

17. *Id.* at 9.

18. *Id.* at 10.

19. See Chhatrapati Singh, "Legal Policy For The Control of Environmental Pollution", in Paras Diwan, (ed.), Environment Protection, 49 at 55 (1987).

20. *Supra* note 11 at 10. The meeting recommended a single system of environmental courts invested with jurisdiction under the Environment Protection Act (and other related environmental Acts), over both criminal prosecution and civil claims for violation of the laws. It was recommended that each state and Union territory should have environmental court of first instance with single environmental appellate court with its headquarters in New Delhi. The court should have the status of a High Court, with right to appeal under article 136 of the Constitution to Supreme Court of India. It was suggested that the Law Commission of India which is charged with priority tasks of judicial reforms in India, should assist the Central Government in a detailed design of environmental courts.

21. A.I.R. 1987 S.C. 965.

22. *Id.* at 982.

23. The meeting of Experts has recommended that in order to realise fully the deterrent potential of this provision, number of rules should be framed by the Central government. See supra note 11 at 45. Section 6 authorises the Central Government to make rules to regulate environmental pollution.

24. Section 7.

25. Section 8.

26. Section 9(1).

27. Section 9 (2).

28. Section 9 (3).

29. Section 10.

30. Section 11.

31. Section 12.

32. Section 13.

33. Section 15 (1) (emphasis is of the author).

34. Section 15 (2).

35. See *supra* note 11 at 41-42.

36. Section 16 (1).

37. Section 16 (2).

38. Proviso to Section 16 (1).

39. See V.S. Chitnis, "The Environment (Protection) Act, 1986: A Critique " in Paras Dewan, *Environment Protection 152* at 156 (1987) For comparative

penalties see sections 41 to 45 of the Water (prevention and Control of Pollution) Act, 19981.

40. See for example. *The Child Labour Act, 1986; The Juvenille Justice Act, 1986; The Consumer Protection Act, 1986.*

41. See *Supra* note 11 at 38-40. See also P. Leela Krishna, "Public participation in Environment Decision making" in *Law and Environment 187-200* (1984).

PUBLIC INTEREST LITIGATION, PROGRAMME AND PUBLICITY

I.C. Saxena

The explosion of population, the expansion of industry with application of science and technology, the easing out of afforestation slowly and slowly in a way leading to deforestation, the delimitation of water sources, the human habits and nature have posed a great danger to the ecological balance. The Bhopal gas tragedy has brought the question of air pollution to the forefront.

Consequent upon the declaration adopted at the United Nations Conference on the Human Environment in 1972, a legislative consciousness has arisen as follows :

1. Water (Prevention and Control of Pollution) Act, 1974.
2. Air (Prevention and Control of Pollution) Act, 1981.
3. Water (Prevention and Control of Pollution) Cess Act, 1977.
4. Environment (Protection) Act, 1986.

Obviously, as stated in the preamble to the Environment (Protection) Act, 1986 this legislation has been made to implement the decisions of the aforesaid conference. Special mention must be made of article 48-A (Directive Principle of State Policy) and article 51-A (g) and (j) (Fundamental Duties) of the Constitution.

Article 48-A : The State shall endeavour to protect and improve the environment and to safeguard the forests and wild-life of the country.

Article 51-A : It shall be the duty of every citizen of India : (g) to protect and improve the national environment including forests, lakes, rivers and wild-life, and to have compassion for living creatures.

(j) to strive towards excellence in all spheres of individual and collective activity so that the Nation constantly rises to higher levels of endeavours and achievement.

The United Nations Conference created the consciousness for a need for environmental protection among the member States and the latter, in turn, must create awareness among its citizens. This is achieved not by mere passage of legislation but by enforcement and implementation of it. Furthermore, the ecological message must be carried to the masses through governmental and non-governmental agencies, voluntary organisations, social reformers, press etc., along with strong judicial attitudes to protect the human environment. A crusade must be fought against pollution of air, water and noise. The following scheme of protection of the environment is listed through :

1. The public interest litigation;
2. The educational and training programme; and
3. The publicity

THE PUBLIC INTEREST LITIGATION

The public interest litigation or social action group or social interest litigation or social justice litigation or social means an action which may be initiated by a public-spirited person, a social reformer, a voluntary organisation or a group for the enforcement of the right of the ignorant, the destitute, the disadvantaged, the disabled on humanitarian grounds in which the mover has no mere personal interest. A letter addressed by such an entity to the Chief Justice or the court or a single judge is accepted as an application for a Writ Petition under article 32 of the Constitution and a show-cause notice is issued to the other party.

In the recent case of *Rakesh Chandra v. State of Bihar*[1], (reported in February 1989), two residents of Patna addressed a letter to the Chief Justice of the Supreme Court regarding the conditions of the inmates of the Mental Hospital at Kanke near Ranchi. The Court treated it as a public interest litigation and issued notice to the State. The Chief Judicial Magistrate had submitted a report to the court which stated that "there were acute shortage of water in the Hospital".[2] There were " five ordinary wells but there was no motor pump installed in any one of them ".[3] Also, "the sanitary fittings were not operating having got choked. The patients were, therefore, forced to ease themselves in the adjacent open field. Consequently the environments had become polluted and unhygienic."[4] The Supreme Court appointed a Committee of management to look after the institution. It opined that, "Parities including the Committee shall have liberty to move this Court from time to time."[5]

There have been four public interest litigation cases by the name of *M.C. Mehta V. Union of India.*[6] Mr. M.C. Mehta is "an active social worker."[7] In the first of these cases (1987), concerning leakage of oleum gas, the Supreme Court was confronted with the industrial puzzle that while science and technology were essential ingredients today in the improvement of the quality of human life, there was a hazard in their use and it was not possible to eliminate the risk altogether. The Court, therefore, while laying down conditions for the restart of the unit,[8] impressed Upon the Government of India to evolve a National Policy for location of chemical and other hazardous industries in areas where population is scarce, every care must be taken to see that large human habitation does not grow around them. There should preferably be a green belt of 1 to 5 km width around such hazardous industries.[9]

In the second of these cases, it was held that the letter need not be addressed to the Chief Justice or the Court or all the companion judges; it may be addressed to one individual judge. Nor need the letter be supported by an affidavit.[10]

In the third case (1988), the petitioner complained that he was moved because "neither the government nor the people are giving adequate attention to stop the pollution of the river Ganga".[11] The court ordered the closure of these tanneries which did not take the necessary steps for the treatment of effluents. It did so consciously because "life, health and ecology have greatest importance to the people,"[12] than employment to workers.

In the fourth of these cases (1988), again a case of public interest litigation, the petition was entertained "to enforce the statutory provisions which impose duties on the municipal authorities and the Boards constituted under the Water Act."[13]

There have been decided cases calling upon the Municipal Committees to perform "their statutory functions and duties in respect of preservation of sanitation". In *Ratlam Municipal Committee V.*

Virdhichand[14] the Supreme Court vigorously enforced the Municipality's duty. *L.K. Koolwal V State*[15] is a public interest litigation. Mr. Justice D.L. Mehta, in his Judgement, said that : "Maintenance of health, preservation of the sanitation and environment falls within the purview of Article 21 of the constitution"[16]. The High Court directed "the Municipality to remove the dirt, filth etc., within a period of six months and clean the entire Jaipur City"[17]. In T. *Damodar Rao V. S.D.* Municipal Corporation Hyderabad[18], the Andhra Pradesh High Court held that "environmental pollution and spoliation should also be regarded as amounting to violation of article 21 of the Constitution".[19]

In the *Kookwal* case, the learned Justice referred to article 51-A, and said that the (fundamental) duty to protect and improve the environment created a right in him for the purpose against the State. Thus "right and duty co-exists".[20] Again *Kinkri Devi V. State*[21] is a social action litigation, paying that the mining lease for the exploitation of limestone be quashed. Assistance was taken of an article in a newspaper as to the consequences of the operation for the ecological balance, environment and natural resources. The Court referring to the view of the Supreme Court in R*ural litigation and Entitlement Kendra, Dehra Dun V State of U.P.*[22] quoted it[23]:

> Mining operations in these areas have led to cutting down of the forest. Digging of limestone and allowing the waste to roll down or carried down by rain water to the lower levels has affected the villages as also the agricultural land located below the hills. The naturally formed streams have been blocked. Blasting has disturbed the natural quite, has shaken the soil, loosened the rocky structures and disturbed the entire ecology of the area.

The Himachal Pradesh High Court appointed a Committee to report in the matter, meanwhile necessary directions were issued in the interest of environmental protection and ecological preservation.

It need be stressed that the apex court has recorded appreciation of the petitioners in public interest litigation and awarded cost. To quote[24]:

> We must place on record our appreciation of the steps taken by the Rural litigation and Entitlement Kendra. But for this move, all that has happened perhaps may not have come. Preservation of the environment and keeping the ecological balance unaffected is a task which not only Governments but also every citizen must undertake.

The Court assessed and awarded Rs. 10,000/- as cost to the Kendra. It is heartening to note that under Section 4 of the legal Services Authorities Act, 1987, the central Authority, subject to the general directions of the Central Government, may start a social justice litigation, *Inter alia*, for environmental protection. Thus the evaluation of development to the public interest litigation in the area of preservation of ecological balance is a pointer to the two strong conclusions : (1) that there is a consciousness among the individuals to preserve it, (2) that the Courts have exhibited a healthy judicial attitudes in the direction. Thus the public consciousness coupled with the judicial approach are fulfilling ecological dreams.

PROGRAMME

The United Nations Conference declaration on Human Environment in 1972 created global consciousness on protection of environment. Ministries and departments of environment began to be created in several countries. India followed suit with a Ministry of Environment at the Centre and environmental departments or cells in the states, with constitution of Water Boards. Legislative enactments were made including amendments in the constitution and the framing of the Air (Prevention and control for pollution) Act, 1981.

The NCERT should prescribe lessons on cleanliness and green scenery for primary students. Lessons should be got written specially for this purpose. The importance of preservation of greenery and habit of cleanliness should be stressed among the students of middle and high schools.

A paper on environmental preservation should be added to a course in geography or as an independent paper at the B.A., and B.Com. levels. At the B.Sc. level technical education on air and water pollution should be introduced. In *M.C.Mehta V. Union of India*[25] the Supreme Court emphasized the value of environmental curriculum. A Committee of experts from science, industry, education and law could be set up for framing of such courses and inviting experts to write special lessons and deliver extension lectures on the subject.

In the aforesaid case, the Court said : "Training of teachers who teach this subject by the introduction of short-term courses for such training shall also be considered."[26]

The Bar Council of India should introduce an optional paper on environmental legislation in L.L.B. II or III year. The Diploma in Labour Law and M.B.A. Courses should have an independent paper on population laws, including industrial pollution.

The industrial houses should start training of limited workers in handling and disposal of industrial waste and effluents and safety and emergency services. Legal assistance could be sought from the Central and State authorities under the Legal Services Authorities Act, 1987.

The social reformers, the voluntary social welfare organisations and the Central Authority under the above 1987 Act should not hesitate to file public interest litigation where the need arises. The government and non-government agencies should co-operate in this aspect.

In the M.*C. Mehta case* the Supreme Court suggested that the Kanpur Municipal Committee should construct public latrines and urinals in sufficient number for poor people. No charge should be levied from the poor for these facilities, otherwise they will use open land for the purpose.

In *M.C. Mehta case*, the Supreme Court also suggested a keep-the-city clean week. It said, "During that week all the citizens the members of the executives, members of parliament and the State legislatures, members of judiciary may be requested to cooperate with the local authorities and to take part in celebrations by rendering free personal service. This would surely create a national awareness of the problems faced by the people".[27]

Laws should be strictly enforced. It is gratifying to note that under section 19 of the Environment (protection) Act, 1986, the court shall take cognizance of a complaint filled by(b) any person who has given notice of not less than sixty days in the manner prescribed, of the alleged offence and of his intention to make a complaint, to the Central government or the authority or officer authorised as aforesaid.

The Municipal Committees should provide facilities to the people to convert manual latrines into flush latrines with well laid sewerage system.

There is no right to do business on public roads or footpaths. So these places should remain free from crowdy and noisy scenes.

Tube-wells and hand-pumps should be installed at strategic places.

The tree plantation (van mahotsava)week be observed during rainy season throughout India.

PUBLICITY

Since we are living in an age of propaganda, the above suggested programme should be publicised. This will create environmental awareness among the public and seek their cooperation. Television, radio and news papers should continue to play their role in the improvement of environment around us. Seminars on environment should get low of publicity. Perhaps we have passed the stage of the often quoted observation of the learned Justice in *Krishna Gopal V. State of Madhya Pradesh*[28] that the society is shocked by single murder but not by an ordinary errant atmospheric pollution which is read merely as a news item, unless things go amiss seriously. The programme and publicity must be carried on unless a national habit conductive to environmental preservation and improvement has been formed.

REFERENCES

1. AIR 1989 (Feg) S.C. 348.
2. *Ibid.*, at 349.
3. *Ibid.*, at 349.
4. *Ibid.*, at 349.
5. *Ibid.*, at 356.
6. A.I.R. 1987 S.C. 965 (Leakage of oleum gas, A.I.R. 1987 S.C. 1086 (Compensation to oleum gas leak victims) A.I.R. 1988 S.C. 1037 (discharge of trade effluents of industries into river Ganga) A.I.R. 1988 S.C. 1115 (Pollution of Ganga at Kanpur, the enforcement of provisions). For full names of the parties see the cases concerned.
7. *Ibid.*, at 1038 (1988).
8. Supra note 6, at 978-80 (1987).
9. *Ibid.*, at 981.
10. Supra note 6 at 1090 (1987).
11. *Ibid.*, at 1038 (1988).

12. *Ibid.*, at 1048 (1988).
13. Supra note 6 at 1126 (1988). In this case, the Court also listed a programme of environmental education and cleanliness at pages 1126-27.
14. A.I.R. 1980 S.C. 1622 (concerning section 133 of the Criminal procedure Code).
15. A.I.R. 1988 Raj. 2.
16. *Ibid.*, at 4.
17. *Ibid.*, at 6.
18. A.I.R. 1987 A.P. 171.
19. *Ibid.*, at 181.
20. Supra note 15 at 3.
21. A.I.R. 1988 H.P.4.
22. A.I.R. 1987 S.C. 359.
23. Supra note 21 at 7.
24. Supra note 22 at 364.
25. Supra note 5 at 1126 (1988).
26. *Ibid.*,
27. *Ibid.*, at 1128 (1988).
28. (1986) G.L.J. 396 at 400. Text may be read in full in the judgement of the Court.

THE ENVIRONMENTAL PROBLEM AND THE RECENT JUDICIAL ACTIVISM

A.K. Sharma

Today society's interaction with nature is so extensive that the environmental question has assumed proportion affecting the entire humanity. Industrialization, urbanisation, population explosion, over exploitation of resources, depletion of traditional sources of energy and raw materials and search for new sources of energy and raw materials, the disruption of ecological balance, the destruction of a multitude of animal and plant species for economic reasons and some times for no good reasons at all are factors which have contributed to environmental deterioration. While the scientific and technological progress of man has invested with him immense power over nature, it has also resulted in the unthinking use of the power, encroaching endlessly on nature. If man is able to transform deserts into oases, he is also leaving behind deserts in the place of oases.

In the last century, a great German philosopher warned mankind, "Let us not however, flatter ourselves much on account of our human victories over nature, for each such victories nature takes its revange on us".

The International association for the protection of Nature and Natural resources calculates that now, on an average one species or subspecies is lost every year. It is said that approximately 1,000 bird and animal species are facing extinction at present. So it is that the environmental question has become urgent and it has to be properly understood are squarely met by man.[1]

The above paragraphs show the anxiety of the judiciary to stretch its hands as far as possible to protect the vital human right to a clean environment, not only this much in total, judicial action supported public interest litigation seems to provide better prevention in environmental matters.

Justice Ranganath Mishra while delivering the judgement in the *Rural litigation and Entitlement Kendra V. State of U.P.* observed that "Prevention of the environment and keeping the ecological balance unaffected is a task, which not only governments but also every citizen must undertake. It is his fundamenatal duty as enshrined in Article 51-A (g) of the constitution.[2]

In the same case the learned judge observed: "It is for the government and the nation and not for the court to decide whether the deposits should be exploited at the cost of ecology and environmental considerations. Government should take a policy decision and firmly implement the same."[3]

On July 14, 1983 a letter received from the Rural litigation and Entitlement Kendra Dehra Dun was directed to be registered as a writ petition under Article 32 of the constitution and notice was ordered to the State of U.P. and the Collector of Dehra Dun. Allegation of unauthorised and illegal mining in the Mussoorie-Dehra Dun Belt which adversely affected the ecology of the area and led to environmental disorder - such direction definitely shows the sensitivity of our judiciary towards the problem.

Our Apex Court has further loosened the fist of procedural technicality in cases of grave public importance like environmental disorder.

Justice Mishra in the above matter observed : "It could not be said that for public interest litigation procedural laws do not apply. At the same time it has to be remembered that every technicality in the procedural law is not available as a defence when a matter of grave public importance is for consideration before the court". Even if it is said that there was a final order, in a dispute regarding closure of mines causing environmental disorder in the hill areas, the plea of *res-judicata* could not be entertained in a subsequent public interest litigation to protect the environment.[4]

In *Sachindanand Randey V. State of West Bengal*[5] Justice Chinnappa Reddy held "Whenever a problem of ecology is before the court, the court is bound to bear in mind Article 48-A of the constitution. The directive parinciple which enjoins that the state shall endeavour to protect and improve the environment and to safeguard the forests and wildlife of the country. Airticle 51-A(g) which proclaims it to be the fundamental duty of every citizen of India to protect and improve the natural environment including forests, lakes, rivers and wildlife and to have compassions for living creatures". When the court is called upon to give effect to the Directive Principles and the Fundamental Rights, the court is not to shrug its shoulders and say that priorities area a matter of policy and so it is a matter for the policy making authority . . . The court may always give necessary directions.

The court may interfere in order to prevent a likelihood of prejudice to the public.

The judicial attempts to ensure the satisfactory protection to the environment were given a further boost in 1988 by Justice Venkataramiah while delivering judgment in writ petition.[6] Justice Venkataramiah held: "The Financial capacity of the tanneries should be considered as irrelevant while requiring them to establish Primary Treatment Plants. Just like an industry, which cannot pay minimum wages to its workers cannot be allowed to exist. A tannery which cannot set up a primary treatment plant cannot be permitted to continue to be in existence for the adverse effects on the public at large.[7]

On the financial aspect the same judge observed: "The crucial question is not whether developing countries can afford such measures for the control of water pollution, but it is whether they can afford to neglect them".[8]

Our Apex court has expressed its satisfaction over the Legislative bulk available in the field but has not spared High Courts and Executive authorities for not assessing the problem as it should be. Justice Venkataramiah directing the High Court held: "Since the problem of pollution of the water in the river Ganga has become very acute the High Courts should not ordinarily grant orders of stay of criminal proceedings in case under Section 482 Cr. P.C., and even if such an order of stay is made in an extrarodinary case the High Courts should dispose of the case within a short period, say about two months from the date of the institution of such case.[9]

In the same case the judge showed his dissatisfaction towards the executive authorities and held: ".... It is unforunate that although Parliament and the State Legislatures have enacted the aforesaid laws imposing duties on the central and state boards and municipalities for prevention and control of pollution of water, many of these provisions have just remained on paper without any adequate action being taken.

The enlightened approach of the Supreme Court towards the environmental problem is evident from the above judgements.

REFERENCES

1. Shri Sachidanand Pandey and another V the State of West Bengal and others - AIR 1987 SC 1109.
2. AIR 1987 SC 359.
3. AIR 1987 SC 363.
4. Rural litigation and Entitlement Kendra V. State of U.P. - AIR 1988 SC 2187.
5. AIR 1987 SC 1109.
6. M.C. Mehta V Union of India AIR 1988 SC 1037 and AIR 1988 SC 1115.
7. AIR 1988 SC 1037.
8. AIR 1988 SC 1115.
9. AIR 1988 SC1115.

LEGISLATIVE AND JUDICIAL EFFORTS TOWARDS ENVIRONMENTAL PROTECTION

Brijendra Kumar Galchhania

A livable environment is the utmost responsibility of man in his own interest. Yet, the natural environment is being severe every day. Man is responsible for a wide range of changes in the ecosystem. Through ill-devised plans, population explosion, increased urbanisation, thoughtless destruction of the forests, adoption of industrial policies that result in large quantities of effluents. Expansion of Science and technology may be said to be the basic cause responsible for the deterioration of the environment. There are no two opinion that environmental degradation should be checked so as to avoid serious and sometimes irreversible damage to life.

The roots of environmental protection can be traced in ancient Indian thinking. Considering rivers as pious, worshipping trees and Sun is a reflection of this philosophy; but inspite of these things, laws concerning environmental protection are only of recent origin. The objective of the laws on environmental protection is to preserve and protect the nature's gift to human being. In India, some legislative, judicial executive and public efforts have been made to achieve the objective of environmental protection. In this paper, an account of goals achieved in legislative and judicial direction is being presented.Under legislative efforts before 1972 Stockholm conference on human environment only some specified provisions related to environmental protection were available in different Acts.

In Indian Penal Code, 1860, provisions regarding negligent and malignant act likely to spread infection of disease, punishment for public nuisance and mischief by fire or explosive substances have been made under Sections 269, 270, 295 and 435.

The *Easement Act-1882*, Sections 7, 15, 28 (d) make provisions for restrictions by law in force on exclusive easementary rights, acquisition of easement by prescription and prescription right to pollute air or water.

In *Indian Factories Act-1948,* Section 12, 92 make the provisions for the disposal of waste material and general penalty for offences.

Doctrine of strict liability for certain acts in common law, based on *Rayland V. Fletcher,* 1868 LR 3H L 330.

Some specific statues have been made in *Atomic Energy Act-1962, Forest Act-1952,* State Smoke Nuisance Acts and Rules for licensing factories.

In June, 1972, the U.N. Conference held at Stockholm on human environment focussed the attention of all nations including India, on the growing world vide menace of environmental contamination arising out of air, water, land and noise pollution. A synthesis of thoughts started on environmental protection in India after Stockholm Conference. It resulted two long waited important acts on the path of intended journey of environmental protection, the Wild Life (Protection) Act 1972 and Water (Prevention and Control of Pollution) Act 1974.

In *Wild Life (Protection) Act-1972,* safeguards for wild life are prescribed to protect and survive innocent disappearing wild animals. In *Water (Prevention and Control of Pollution) Act-1974,* provisions regarding prohibition and penalty on the use of stream or well for disposal of polluting matter, restriction on new outlets and new discharges and existing discharge of sewage of trade effluent are made under section 24 to 26 and 42 to 45.

Composition of Central, State and joint Boards are prescribed for prevention and control of water pollution under sections 3 to 18. The State Government are empowered to make rules and to use discretion to apply this act to certain areas and provisions for cognizance of offences, water laboratories etc. are legislated in this act.

During Constitution Amending era some milestone steps were taken on the path of environmental legislation. It was in the *Constitutional (42 Amendment) Act-1976* that for the first time Indian legislators felt the paramount need of protecting and preserving environment from pollution.

This act inserted Article 48 A in the *Directive Principles of State Policy*. Article 48 A says "The State shall endeavour to protect and improve the environment and to safeguard the forests and wild life of the country." A new chapter 4 A was also inserted on Fundamental Duties which provided amongst others the duty of every citizen of India to protect and improve the natural environment including forests, lakes, rivers and wildlife and to have compassing for living creatures.

In continuation of above, Forest Conservation Act-1980 and the Air (Prevention and Control of Pollution) Act-1982 legislated to control environmental pollution and other health hazards. *Forest Conservation Act-1980* contains the provisions for prevention of woods and implementing machinery frame work for conservation. In *Air (Prevention and Control of Pollution) Act-1982*, the following important provisions are made:

State Government are empowered to declare air pollution control areas, to give instructions for ensuring standards for emission from automobiles, to restrict use of certain industrial plants, to prohibit certain industries not to allow emission of air pollutants in excess of the standards laid down by State Board (under Sections 19 to 22) and penalties for contravention of above provisions and failure to comply provisions of this act, rules and orders (Under Sections 37 to 39).

It is proposed that existing Central Board and State Boards for the prevention and Control of Water Pollution, will also perform the functions of prevention and Control of Air Pollution under Section (3-18) Provisions for inspection, Laboratories, cognizance of offences are also provided.

It was the Bhopal MIC gas tragedy the 'Chemical Hiroshima' of 1984 which made the people aware of ill effects of pollution and catalysed some important legislation to control it. To cope with the need of the time Environment (Protection) Act-1986 is legislated, existing Indian Factories Act is amended and Environment (Protection) Rules-1987 are framed.

Factories (Amending) Act-1987 contains the following provisions from environmental and workers safety point of view :

'Hazardous Process' is defined in Section 2 (cb) as any process or activity in an industry causing injury to the workers or resulting in environmental pollution if special care is not taken. Provisions relating to hazardous process like specific responsibility of the occupier, permissible limits of exposure of chemicals and toxic substances, emergency standards, are made in new chapter IV-A under Section 41 C, 41 E, 41 F. Duties of occupier, duties of manufacturer, provisions for 'Site Appraisal Committees' with extended power of inspectors are added under Sections 7A, 7B, 41A and 9.

The Environment (Protection) Act-1986 is legislated as an urgent need for the enactment of a general legislation on environmental protection, because the existing laws–generally focus on specific pollution or hazardous substances. It was further aimed for coordination of activities of the various regulating agencies, creation of authorities with adequate powers for environmental protection, speedy response in an event of accident. Some important provisions of the Act are as follows :

Central Government is expowered to take measures to protect and improve environment, to appoint officers and prescribing their powers and functions, to give directions and to make rules for regulation of environmental pollution under Sections 3 to 6 and 25. For prevention, control and abatement of environmental pollution, Act embodies, not to allow emission or discharge of pollutants in excess of the standards, to handle hazardous substance taking procedural safeguards and regarding penalties for contravention of the provisions of the Act, rules, orders and directions and cognizance of offence under Sections 7, 8, 15, and 19. Act defines under Section 2 (a), (b), (c).

'Environment' includes water, air, land and their inter-relationship with human beings other living creatures, plants, micro-organism and property; *'Environmental Pollutant'* as any solid, liquid or gaseous substance injurious to environment; and *'Environmental Pollution'* as presence of any pollutant in the environment.

Apart from the legislation, the efforts of the Indian judiciary in environment pollution control are indeed laudable, particularly when the legislature is legging behind in bridging the lacuna in the existing legal system. The Supreme Court and High Courts laid town important legal doctrines and expressed a new, even revolutionary meaning to the constitutional provisions to protect environment through social vision.

In *Ratlam Municipality V. Vardhi Chand*, AIR 1980 SCC 1622, Supreme Court highlighted on the dangers of dirt, squalor, stench and stink due to open drains and public excretion by humans to human environment and health hazards accountability of statutory or governmental bodies irrespective of financial implications.

In *Rural Litigation and Entitlement Kendra, Dehradoon V. State of U.P.*, and *Devki Nandan Pandey V. Union of India*, AIR 1985 SCC 652, in writ petition under Article 32, Supreme Court ordered the closure of certain Mussoorie Hill lime stone quarries causing imbalance to ecology and hazard to public health.

In *M.C. Mehta Vs. Union of India*, AIR 1986 SC 176 Supreme Court directed the 'Shriram Food and Fertilizer Company' for leakage of chlorine gas from their plant, to take all necessary safety measures before reopening the plant, to deposit a sum of Rs. 20 lacs and to produce a bank guarantee for Rs. 15 lacs as a security for the compensation claims of the victims of oleum gas leak.

In the same case on a reference, and *Sriram Food and Fertilizer Industries V. Union of India* AIR 1987 SCC 965, Supreme Court laid down these important principles :

(i) The Oppressed, exploited poor in India can seek enforcement of their fundamental rights[1] from the Supreme Court by writing simple letter to any judge (even not accompanied by an affidavit)

(ii) If an enterprise, an accident occurs in the operation of hazardous or inherently dangerous activity resulting in escape of toxic gas, the enterprise is strictly and absolutely liable to compensate all those who are affected by the accident and such liability is not subject to any of the exemptions of Strict Tortuous liability[2] under the *doctrine of Rayland V. Fletcher* (1868) LR 3HL 330.

(iii) The Court also pronounced that the larger and more prosperous the enterprises the greater must be the amount of compensation payable to an accident victim.

Although Supreme Court has not referred *'Article 21'* in these judgements but it is implied that the court entertained environmental complaints involving violation of Article 21's *'Right to life.'*

In *T.Damodar Rao V. S.O. Municipal Corporation of Hyderabad* AIR 1987 A.P. 171 High Court clearly made a bold pronouncement that:

"Protection of the environment is not only the duty of the citizen but it is also the obligation of the State and all State organs including courts." It is the legitimate duty of the court as the enforcing organ of constitutional objectives to forbid all actions of the State and the citizen from upsetting the environmental balance. In this case court held that the slow poisoning[3] by the polluted atmosphere caused by *environmental pollution and soilation should also be regarded as amounting to violation of Art. 21 of the Constitution.*

In *Sachidanand Pandey V. State of W.B.* AIR 1987 SC 295, the Supreme Court held that court may examine[4] the Government decision to allot land for the construction of a five star hotel near by the zoological garden of Calcutta, and its impact on the ecological balance and the 'bird migration.'

In *Anil Krishna Pal V. State of W.B.* AIR 1989 Cal. 102, High Court held that if the public body like municipality refuse to take steps on accordance with law to remove the grievances of a citizen, a citizen can invoke writ jurisdiction of court for appropriate relief.[5]

In *'Bhopal Case'*, Supreme Court really functioning as a 'Lok Adalat' catalysed and ordered a settlement between the Government of India and the Union Carbide Company, quashing all present and future suits against Carbide. Although Carbide has paid 705 million rupees or 47 million dollars for the compensation of the Bhopal Gas victims, but I never thought that all cases against Union Carbide would be quashed. In this Bhopal Gas case, which had travelled from New York to Supreme Court through Bhopal District Court and M.P. High Court apart from all controversies,[6] the Supreme Court laid down that *'Multinational Companies' will now be fully liable for the actions of overseas subsidiaries.*

CONCLUSIONS AND SUGGESTIONS

It is proposed that *more efficient delegated legislation from Gram Panchayat level to State level is required. More strict laws, by laws to penalise polluters, provisions for reward and their disclosure to public through 'Mass-Communication Media' for awareness are needed.*

For judicial efforts, it is clear from the above that Indian judiciary for protecting environment, is functioning in harmony with our constitutional goals, acting as bridging device between legislation and constitutional objectives. Because our Apex court and High courts are over burdened with lot of cases. I, therefore suggest that for successful administration of environmental law, *construction of "Environmental Tribunals" and delegation for some quasi-judicial-powers to appropriate authorities are needed.*

REFERENCES

1. Through Public Interest Litigation Writ under Article 32 in Supreme Court and 226, 227 in High Court.
2. The court made a boastful assertion in observing: "We have to evolve new principles and lay down new norms which would adequately deal with the new problems which arise in a highly industrialized economy. We can not allow out judicial thinking to be constricted by reference to the law as developed in England recognises certain limitations and responsibilities." AIR 1986 SCC 176.
3. It is accepted that pollution "is a slow agent of death and if it is continued the next 30 years as it has been for the last 30, it could become lethal." Justice Krishna Tyer in 'Pollution and law'.
4. Court held that in view of the Article 48 A and 51 A (g) whenever a problem of ecology is brought before the court it would not refuse to interfere only on the ground that priorities are matter of policy. AIR 1987 SCC 295.
5. The citizens are ordinarily entitled to appropriate relief, once it is shown that their rights have been illegally or unconstitutionally violated. AIR 1989 Cal. 102.
6. The Supreme Court order dated 14-15 Feb., 1989 occasioned howls of dismay. Former Chief Justice of India, P.N. Bhagwati criticised "The multinational has won and the people of India have lost." Many questions are raised, few important are these:

   How did the court transform and appeal against the high courts interim relief order into final settlement?

   How did the court ordered to quash all present and future suits against Carbide?

   How did the court arrive at the figure of 470 million dollars?

PESTICIDE PETTIFOGGERY

S.K. Agarwal

It is often said, and rightly so that war is too important a business to be left to the generals. By the same analogy, there is absolutely nothing for us to dwell in fear of the situation where it can be said that judiciary is too important an institution to be left to the sole care of the judges. Much water has flown down the bridge since the time when justice as administered by the courts was considered to be a cloistered virtue - something esoteric meant to be lopped up as a gospel truth and not to be debated and deliberated upon by the people at large. The judges being human beings cannot remain unaffected and uninfluenced by what is happening around them. The great tides and currents, so says a great American judge, which engulf the rest of the mankind cannot turn aside and pass the judges by.

SETTLEMENT

The hullabaloo arising from the judgement which the constitution Bench presided over by the Chief Justice Shri R.S. Pathak of the Supreme Court has handed down on February 14, 1989, in the Bhopal gas leak tragedy case awarding the payment of U.S. $470 million (Approx. Rs. 705 crore) by the Union Carbide Corporation (UCC) to the victims of the tragedy in full and final settlement of their claims, past, present and future, and at the same time discharging the defendants from all liabilities and responsibilities, both civil and criminal, is understandable. The reasons are not far to seek. First, the settlement came about with such suddenness and surprise as the MIC gas leak from the Bhopal plant on the fateful night of December 2-3, 1984. The Supreme Court of India succeeded where judge Keenan of the southern district of New York and judge M.W. Deo of District Court, Bhopal had failed. The court spoke of the "pressing urgency to provide substantial and immediate relief to the victims. We consider the case predominantly fit for overall settlement covering all litigation of all claims and rights and liabilities arising out of the disaster and accordingly hold it just, equitable and reasonable to order and we do order". Second, it concerned mass industrial disaster of gigantic proportion which left about 3300 persons dead nearly 30,000 persons permanently disabled and incapacitated. Around 60,000 people continue to be adversely affected by inhalation of MIC. A survey done by Tata Institute of Social Science has shown that at least 1,00,000 families have suffered, a majority of them belonging to lower income groups. Third, the catastrophe in question was the direct result of several acts of omission and commission of UCC. It appeared to be case of planned mayhem. Fourth, the government of India put on the statue book, the Bhopal gas leak Disaster (Processing of Claims) Act, 1985 to deal with the claims on behalf of the victims in an appropriate and expeditious manner. Fifth, it threw up many noval and extraordinary questions of fact and law in the legal circles in both India and USA. It was for the first time that the Union of India, acting parents - patriae, took up the cudgels on behalf of the victims against a giant multinational. Sixth, the damage done to the victims could not be compensated in terms of money. The government of India claimed 3.3 million from UCC. It was by all standards an astronomical figure. Last but not least, it was an unequal fight between thousands of claimants on the one hand and a giant multinational on the other, with the later determined to delay and defeat the justice due to the former through a protracted litigative battle (Bansal, 1989).

HISTORICAL BETRAYAL

A trial in the court is atleast a forum for an open fight. It can draw forth the hidden facts, and clarify how rootless and inadequate the assailants statements are. Experience of the Morinaga Arsenic Milk Poisoning and Minamata Mercury poisoning clearly demonstrate that Third party organisations created by authorities have invariably stood on the side that represses the victims. The committee for compensating Minamata victims (Chigusa Committee) and its associated Central Pollution Adjustment Committee (Chue Kogai to Chosei iinkai) which acted as moderators in the Minamata case, were totally useless in relieving the victims.

Between the victims and assailants of pollution there exists several basic discrepancies. The most fundamental is that the two have completely different standpoints in understanding pollution itself. Since, the victim suffers the ravages of pollution with their entire body, there is a total comprehension which includes aspects that cannot be expressed in words and numbers, The assailants understanding of pollution, on the other hand, is limited at best to objectively expressible words and numbers. This basic discrepancy cannot be bridged by a third party. So long as one fails to totally comprehend pollution from the victims stand point, one cannot understand it (Ui, 1974).

The Bhopal Gas Leak (Registration and Processing Claims) Act, 1985 appears like a system of Pollution Insurance. Under the umbrella of recent judgment of our apex court, the multinational UCC could avoid the troublesome disputes which might hurt the social reputation of the company, and at the same time saved the assailants from criminal penalties.

The critics of the judgement and, for that matter, the settlement put it down as historical betrayal. According to them, the Indian Supreme Court has set a very dangerous precedent for the potential victims of industrial accidents in the developing countries. It has aborted all litigation arising from the Bhopal accident of December 1984 and settled all claims for the sum of only US $470 million (Sharma, 1989).

The 5,70,000 claimants for the compensation will, on an average, get less than US $ 600. This figure is reached after deducting the US $ 100 million already spent by the government plus the fees of the private American lawyers and accumulated medical expenses.

In the process, UCC has been relieved of the fear that earlier US precedents, might be repeated for Bhopal. The Manville Corporation paid US $ 2.5 billion to 60,000 people who had suffered lung damage from asbestos, and A.H. Robins and Co. had to pay US $ 250 million to only 9,450 victims of industrial poisoning.

"The government used the judiciary to disguise what was really a settlement. Unfortunately, the court fell a victim to this strategy", says the former Chief Justice of Indian Supreme Court Mr. P.N. Bhawati. Mr. Bhagwati is clearly of the opinion that it was not a court judgement but a settlement between the Indian government and the UCC. The government seems unwilling to accept responsibility for striking a deal which goes against the Bhopal victims, so the settlement is being portrayed as a Supreme Court judgement. It appears that this antipollution agreement has been made with an eye to luring multinational enterprises in the country.

However, doubts are being raised about whether or not the court is exceeding its judicial jurisdiction by decreasing such a wide ranging settlement and quashing all present and future suits against UCC. Mr. Bhagwati wonders, "Why were these offers (of US $ 470 million) and counteroffers made in the chamber of the Chief Justice and not in an open court ?"

It was not money alone that was at stake in the Bhopal case. To avoid future Bhopals, it was necessary to hold UCC liable for its criminal negligence and violations of all safety procedures at the Bhopal pesticide plant.

"By not pinning the multinational down to accepting culpability, the Indian government has left the door open for such companies to persist in diluting their safety standards in Third World countries, "comments the Indian Express in New Delhi.

"In any event, the victim groups stand entitled to seek a review of the final order of the court", says Upendra Baxi, a jurist in new Delhi. According to Mr. Baxi any order that the Supreme Court may finally pass down will not prevent the people from bringing individual actions against the government of Madhya Pradesh as well as the government of India.

"We will file 10,000 individual application before the Supreme Court seeking a review of the decision," says Abdul Jabbar Khan of Bhopal's Gas affected Women's Association. In addition there is a petition pending before the Supreme Court challenging the constitutional validity of the Bhopal Gas leak (Registration and Processing Claims) Act, 1985. If it succeeds, the victims will be entitled to receive their individual actions and the government's *locus-standi* would be gone.

On hearing the news of the settlement the victims of the disaster protested against the woefully low compensation and accused the Indian government for "selling down" to the multinational UCC. Having failed to restrict liability to its poorer Indian subsidiary the, multinational reportedly indulged in disposing of its assets so as to avoid just payment. Later, the company took advantage of every legal stratagen, pressure tactic an dilatory device it could think to drag out the proceedings and frustrate the victims.

With this unreasonable settlement UCC estimates that the impact of the case will not exceed 50 US cents (approximately rupees eight) per share, and will be charged against 1988 reported earnings. Already a sudden spurt in UCC share on the New York Stock Exchange has been witnessed.

HISTORICAL ACHIEVEMENT

The supporters of the settlement hold it up as a full-bodied product of the most judicious and sympathetic endeavours of the apex court. It appears that the atmosphere is surcharged with emotion. In order to have a dispassionate view or the issues involved it is necessary to give here brief resume of the facts concerning legal proceedings which found their culmination in the recent settlement.

FACTUAL MATRIX

The historical disaster which struck Bhopal and its unfortunate inhabitants produced agonised echoes throughout the world, especially the Third World countries. Its reverberations were heard very loudly in the American homes with the result that hundreds of attorneys from different states of USA looked to Bhopal to get from the victims the relevant papers on the basis of which they could file the claims on their behalf in the American courts. As a sequel of this, hundreds of claims petition were filed in 50 states of America. All these claims petitions were later on transferred to the file of the judge Keenan who presided over the southern district court of New York City.

In the meantime, the government of India enacted the foregoing Act of 1985 which vested in it the exclusive right to claim compensation from UCC on behalf of the victims. In exercise of the power vested in it, the government preferred the claims for compensation in the court of judge Keenan. The position that the UCC took-up was that all the claims were to be heard and decided by the courts in India in as much as the Indian courts constituted the proper and convenient forum to decide them. The government of India contested it on various grounds and wanted the American court to decide the controversy. In the first place, the United Carbide India Limited (UCIL), in which the UCC held 50.9% of the shares, would not be in a position to satisfy the claims beyond US $ 95.3 million which represented the total assets of UCIL in India. Thus, even if the decree was passed against the UCC it would remain unsatisfied. In the second place, UCC never expressed its readiness to satisfy the decree which might be passed against it by the Indian courts. It could contest it in the American courts on one ground or the other, thereby delaying and defeating the claims of the victims. In the third place, the Indian courts had not yet dealt with the claims arising from such a mass disaster as the Bhopal gas leak tragedy. Lastly, extreme delays are a historical fact of life in the

Indian courts where even routine and simple proceedings become protracted *oddsseys* and it is not unusual for legal proceedings to long survive the litigants themselves.

FORUM NON-CONVENIENTS

Judge Keenan held the marathon sitting and ultimately on May 5, 1986 ordered the transfer of all cases to Bhopal district court on the ground that Bhopal court constituted the most convenient forum to thresh out all the questions of fact and law. This he did by relying upon the doctrine of forum non-convenients.

INTERIM RELIEF

The scene of the battle shifted to district court Bhopal. The matter proceeded there at snails pace. On April 1, 1987 Judge M.W. Deo of district court Bhopal asked UCC to pay appropriate interim relief to the victims. On August 17, 1987 UCC proposed to pay 46 lakh dollar as interim relief to the victims. Government of India did not agree to this proposal on the grounds that the amount was very meagre. On November 27, 1987 judge Deo heard the appeals on interim relief, and on December 17, 1987 he ordered the payment of US $ 27 crore as the interim relief by UCC to the victims. On January 17, 1988 UCC went to the Madhya Pradesh High Court in appeal. It claimed that the interin relief order was against law. On January 7, 1988 UCC complained that the interin relief payment order of Bhopal district court was not clear, as it has no mention of the names of the people to receive the relief. On April 4, 1988 Justice S.K. Seth of Madhya Pradesh High Court upheld the order of judge M.W. Deo. But justice Seth reduced the amount of interin relief from US $ 27 crore to US $ 19.5 crore, to be paid within a period of two months. This order of justice Seth was called into question before the Supreme Court from both sides i.e. the UCC and the government of India. The legal proceedings in the Supreme Court ended up in the settlement in question.

CRITICISM ANSWERED

The alleged inadequacy of compensation is the thin end of the wedge. The UCC was prepared to give US $ 600 million as compensation. Why did the government of India scale down the figure of US $ 3.3 million, which it initially demanded, to US $ 470 million, which it has now agreed to accept, they ask ? The claimants who number around 6,00,000 will get a paltry sum of money. It can neither compensate the past damage nor suffice the future treatment and rehabilitation of the victims. The criticism ignores sound mathematical calculation.. UCC offered US $ 600 million to be paid over a period of 10 years. It is worth US $ 339 million now at the rate of 12 per cent reduction. On the otherhand, US $ 470 million in 5 years and over one billion dollars in 10 years. The interest on US $ 470 million will be over Rs. 20 lakh a day, Rs. 6.5 crore a month and Rs. 78 crore a year. This amount of compensation has already been paid off.

The case involved serious questions concerning culpability of UCC for the damages. The responsibility for the damage arising from a callous disregard for human life, environment and other safety measures must have been fastened upon an erring multinational. There is no reason, so they argue, why the doctrine of absolute liability which was applied by the Supreme Court in the DCM case for the Oleum gas leak should not hold good in this case also. The critics wittingly or unwittingly pity the plumage but forget the dying bird. They want certain questions of fact and law to be determined at the expense of the victims. They forget that the settlement has been reached on human considerations. The idea seems to be to ensure that the sufferings of the victims do not continue endlessly through prolonged litigation. We should be concerned more with the victims and their cause rather than with the pyrrhic victory which the legal activists may secure in the future.

The discharge of UCC from the criminal liability for its acts of mission and commission is in violation of the constitutional provisions. In a case, so they allege, or mass murder and mayhem. This impropriety is further compounded by the failure on the part of the government to take into confidence the victims of the tragedy. The Supreme Court cannot bar the trial of the criminal cases. This criticism seems to be punctured from the very beginning. The Act of 1985 empowers the government not only to aggregate to itself all the cases concerning the tragedy but also to enter into a compromise with the defendants. It was an extraordinary case and thus required by its very nature exceptional measures to deal with it. The government have the power to withdraw even the pending criminal case from the courts. The victims numbering around 6,00,000 cannot be consulted on an individual basis in a case of this type.

The critics have caused aspersions even upon the court with regard to the manner in which it has stamped the settlement with its judicial imprimatur, that too within half an hour. Perhaps the critics forget that it is the duty of the court, which is seized of the matter, to make sincere endeavours to make the contending parties clinch a just and reasonable settlement of the dispute. We must be beholden of the apex court for the salutary role it has played in the matter. It succeeded where all others had failed. Even assuming, but not admitting, that it was an out-of-the court settlement there was nothing wrong for the court to set its seal of approval over it. Any attempt to review the judgement on the lines of the Antulay case would be a self-defeating measure.

It is high time we steer clear of the miasms of charges and countercharges, proceed in right earnest towards the treatment and rehabilitation of the helpless and helpless victims of the tragedy and ensure that the compensation money reaches the rightful claimants (Bansal, 1989).

REFERENCES

1. Bansal, J.P. 1989. The Bhopal Gas Leak tragedy case - Historical achievements v. Historical betrayal. In *Souvenir District Officers Club,* Kota, pp. 1-5.

2. Sharma, R. 1989. Bhopal settlement - Averaging under US $ 600 per person. *Ecoforum* (Holand), 13 : 6 : 6.

3. Ui, J. 1974. The place of law suits. *Kogai* (Japan), 4 : 8-10.

NEED FOR EFFECTIVE IMPLEMENTATION OF ENVIRONMENTAL LAWS

T. Bhattacharyya

The global problem of deterioration of the environment has been sought to be tackled in India, *inter alia* by the enactment of the following specific legislations on the subject :

1. The Forest Act, 1927.
2. The Forest Conservation Act, 1980.
3. The Water (Prevention and Control of Pollution) Act, 1974.
4. The Air (Prevention and Control of Pollution) Act, 1981.
5. The Environment (Protection) Act, 1986.

Each one of the above mentioned laws tries in its own way to protect and improve, if possible, the environmental damage which has almost reached alarming proportions in our country.

The Forest Act, 1927 is a comprehensive legislation with more than eighty sections in it. It was enacted to keep the ecological balance and protect the forests for the overall benefit of the living organisms. Among other important provisions, the Act provides to preserve forests, appointment of forest settlement officers, acquire land over which right is claimed, order on claims to rights of pastures or to forest produce, appeals, revise arrangement, stop ways and water courses, lay penalties, seize property liable to confiscation, prevent commission of offences etc.

The Forest (Conservation) Act, 1980 is small piece of legislation dealing with restriction on the dereservation of forests or use of forest land for non-forest purpose, constitution of an advisory committee and power to make rules etc.

The Water (Prevention and Control of Pollution) Act, 1974 with certain amendments in 1978 is an extensive legislation with more than sixty sections for the prevention and control of water pollution. Among other things, the Act provides for constitution of Central and State Boards for preventing water pollution, power to take water samples and their analysis, discharge of sewage or trade effluents, appeals, revision, minimum and maximum penalties, publication of names of offenders, offences by companies and Government departments, cognizance of offences, water laboratories, analysis etc.

The Air (Prevention and Control of Pollution) Act, 1981 is also a comprehensive legislation with more than fifty sections. It makes provisions, *inter alia*, for Central and State Boards, power to declare pollution control areas, restrictions on certain industrial units, authority of the Board to limit emission of air pollutants, power of entry, inspection, taking samples and analysis, penalties, offences by companies and Government and cognizance of offences etc.

The Environment (Protection) Act, 1986 is an exhaustive legislation with more than twenty five sections. The word environment has been given a wide meaning by including within it water, air, and land and the inter-relationship which exists among and between water, air and land, and human beings, other living creatures, plants, micro-organisms and property. Similarly, environmental pollution has been defined to mean any solid, liquid or gaseous substance present in such concentration as may be or tend to be, injurious to environment. The Act also provides for power of Central Government to take measures to protect and improve environment, appointment, powers and functions of officers, bar of emission or discharge of environmental pollutants in excess of the standards, procedural safeguards for persons handling hazardous substances, powers of entry, inspection and taking samples, laboratories, analysts, penalties, offences by companies and Government departments and cognizance of offences.

Thus, it is clear that the above mentioned five legislations cover a reasonably wide area. As the time goes by other subjects which need to be brought within these laws will come to be identified. However, in comparison there have been only a very few cases which have gone before the courts and still fewer have gone upto the Supreme Court. All the decisions of the Supreme Court so far have been notable and fully in keeping with the spirit of the laws.

In *Municipal Council, Ratlam v. Virdi Chadand and others,*[1] a complaint under section 133 Code of Criminal Procedure, 1973 was filed before the Sub-Divisional Magistrate by the respondent who were the residents of locality within the Municipal limits of Ratlam. They alleged that the Municipal Council had failed to prevent discharge on public street from a nearby alcohol plant, and that it had also failed to provide roads and sanitary facilities including public conveniences for slum-dwellers who were using the road for that purpose. The appellants took an improper, unfortunate an legally unsound plea that the respondents had gone to reside in the locality with full knowledge of insanitary conditions and polluting discharge from the alcohol manufacturing unit.

The Magistrate dismissed the plea and held that the Municipal Council must provide amenities and construct drain pipes to flush the excreta. He also threatened prosecution under section 188 of Indian Penal Code in case on non-compliance. The District and Sessions Court found the order not justified but the High Court upheld the Magistrates order.

The Supreme Court dismissed the Municipal Councils appeal and held that statutory liability could not be undone because of economic constraints. The Council was ordered to stop the discharge of effluents from the alcohol plant on the street. The Sub-Divisional Magistrate was directed to act under section 133, Code of Criminal Procedure, 1973 to abate the nuisance. It was emphasized that industries could not be allowed to make profit at the expense of public health. Six month's time for construction of enough separate public latrines for men and women was given. It was to be ensured that these stayed clean. The State Government was directed to give special instructions to the Malaria Eradication wing. The Council was further ordered to see that no filth accumulated. The Court pertinently pointed out that extra expenses on these would make consequential cut in the medical budget. In case of wilful breach the Court suggested contempt action.

In *Rural litigation and Entitlement Kendra and others v State of Uttar Pradesh*[2] a letter by the petitioner alleging unauthorised and illegal mining operation carried on in Mussorie hills adversely affected the ecology of the area leading to environmental disturbances was treated as a writ. The Supreme Court stopped all fresh quarrying. Two Committees, one to report into the matter of mining operations and the other to consider ecological and environmental problems with reference to mining, were constituted. The question of carrying on mining operation *Vis-a-vis* its adverse effect on environment along with the plea that limestone quarries of the area fulfilled about three percent requirement of the country were involved.

The Supreme Court asked for indulgence of the Government of India to decide whether the deposits should be exploited at the cost of ecology and environmental considerations. It emphasised the need for maintaining a proper balance between preservation and utilisation of the deposits on the basis of expert advice. The Supreme Court cautioned that the assets were for the whole mankind and were not to be exhausted in one generation - there had to be a long term perspective. Preservation of proper environment was termed by the court as a social obligation and constitutional duty.

M.C. Mehta v Union of India[3] popularly known as the Shriram Foods and Fertilizer Industries case, was a public interest litigation for determining the liability of a large enterprise in a populated area manufacturing the hazardous products. The Supreme Court hoped for reduction of the element of hazard or risk to the community by taking all necessary steps for locating such industries in manner which would pose least risk or danger to the people and maximise the safety requirements. An attempt was made to impress upon the Government of India to evolve a national policy for location of chemical and other hazardous units in scarcely populated areas.

The Supreme Court held that enterprises engaged in hazardous or inherently dangerous activities are strictly and absolutely liable to compensate those who are affected by accidents related to those activities. The concept of punitive damages was emphasised in such cases. It was stressed that larger and more prosperous the company is, higher the amount of compensation and damages payable by it. The enterprise must be held to be under an obligation to provide that the hazardous or inherently hazardous activity in which it is engaged must be conducted with the highest standards of safety and if any harm results on account of such an activity to its workers or residents in its surroundings the enterprise must compensate for such harm. It would be futile for the enterprise to say that it had taken all reasonable precautions and that the harm was caused without any negligence on its part.

In *Shri Sachidanand Pandey and another v. State of West Bangal and others*[4] a plot of land measuring four acres out of the mand of the

Zoological Gardens, Calcutta was allotted to a well known group of hotels on lease by the Government of India for the construction of five star hotel there. The main environmental concern was the likely disturbance to the inhabitants of the zoo, especially the birds, from the glowing lights and heavy vehicular traffic associated with the hotel.

The Supreme Court observed that industrialisation, urbanisation, population explosion, over exploitation of resources, depletion of traditional sources of energy and raw material, the disruption of natural ecological balance, the gains for economic reasons and sometimes for good reason at all are factors which have contributed to environmental deterioration.

In other case of *M.C. Mehta v. Union of India and others*[5] where in the owners of some of the tanneries discharging effluents from their factories in Ganga and not setting up primary treatment plant inspite of being asked to do so for several years. The Supreme Court observed that the river Ganga is the life-line of millions of people of India. Indian culture and civilization has grown around it. The Ganga has always been an integral part of the nations history, culture and environment. The Court issued directions for the closure of those tanneries, though it may bring unemployment, loss of revenue, but life, health and ecology have greater importance to the people.

In *Ambica Quarry v. State of Gujarat*[6] the importance of conservation was stressed by the Supreme Court. The State Government has rejected the renewal of mining licences of certain quarry in Valsad. The Supreme Court held that the Forest Conservation Act 1980 was an Act in recognition of the awareness that deforestation and economic imbalance as a result of the deforestation have become a social menace which should be prevented. The primary duty is towards the community and the obligation to society must predominate over obligation to individuals.

In the famous *Modi Industries Case v. Uttar Pradesh*[7] Pollution Board had sought to prosecute the Modi Industries for discharging untreated effluents into the Kali river. The Judicial Magistrate of Gaziabad upheld the plea. In the Allahabad High Court, however, the industrial house pleaded that only the distillery unit should be prosecuted and not the industrial group, this was accepted.

The Supreme Court, in appeal, ordered the prosecution of the Chairman and eight Directors of the Modi Industries Limited for discharging highly noxious and polluted trade effluents from the Modi distillery into the Kali river. It was very pertinently pointed out that every person who at the time of the commission of the offence was incharge of and responsible to the company for the conduct of business of the company, as well as the company, shall be deemed to be guilty of the offence and shall be liable to be proceeded against and punished accordingly.

These observations are far reaching in content. It is clear that the company in such instances can not escape responsibility for the action of one of its units or subsidiaries. The Uttar Pradesh Pollution Board was also taken to task for delaying in the manner in which it drafted the order against the industry thereby providing it a technical loophole.

A perusal of the above judicial decisions show that the Supreme Court has taken the various laws relating to environmental pollution in the right spirit. It has, in fact, gone far ahead and given judicial lead to all those who care for the conservation of the environment. It has been doing a pioneering work and has not missed any opportunity.

But over-dependence on courts is not desirable. A court can act only when a matter comes before it. This is its inherent limitation. We find that in our country environmental pollution is alarming in every sphere where the question of environment is involved in any way. The need of the hour is not to improve the law as such because that will be taken care of the natural way by passage of time. The emphasis should be on effective implementation of the existing laws, and whenever with passage of time new laws are born, their proper implementation.

It is important to note, however, that the above mentioned far-reaching observations of the apex Court could have been more meaningful if the delay in the final disposal of the cases could be cut. Justice delivery system of India, especially in civil cases, is so slow that it almost defeats the very purpose. It is disturbing to find that each or the above mentioned cases has consumed a long time.

To effectively implement the anti-pollution laws, and especially to cut down the delay, the idea of introducing Environmental Courts has been mooted which would deal exclusively with environmental cases. These Courts could be on a zonal or Regional level, to start with. Since experts in the field of environment will be needed to effectively deal with all kinds of questions relating to ecology and environment, it would be advisable to have at-least two experts to help a professional Judge in delivering a judgement. The system could be a two-tier one and the appeals from the Environmental Courts could be disposed of directly by the Supreme Court. This would very effectively cut down delay and would check attempts to frustrate the very purpose of having environmental protection legislation.

The Bhopal disaster has opened our eyes and therefore, we could have a mechanism to pay some immediate interim mandatory amount of money to those affected by hazardous pollutants. This mechanism could be in the form of public liability insurance of the like. It must also be insured that the ordinary citizen, especially the deprived classes, are not harassed by the rich people and business houses. This can be taken care of by appointing only such persons as officers of the Environmental Courts who have proven credibility and unflinching reputation.

REFERENCES

1. A.I.R. 1980 S.C. 1622.
2. A.I.R. 1987 S.C. 359.
3. A.I.R. 1987 S.C. 965.
4. A.I.R. 1987 S.C. 1109.
5. A.I.R. 1987 S.C. 1037.
6. A.I.R. 1987 S.C. 1073.
7. A.I.R. 1987 S.C. 684.

INDEX

C

D

E

F

G

H

I

J

K

L

M

N

O

P

R

S

T

Y

Z